Marsh Mud and Mummichogs

EVELYN B. SHERR Marsh Mud

The University of Georgia Press ATHENS & LONDON

and Mummichogs

An Intimate Natural History of Coastal Georgia

A Wormsloe
FOUNDATION
nature book

© 2015 by the University of Georgia Press
Athens, Georgia 30602
www.ugapress.org
All rights reserved
Designed by Erin Kirk New
Calligraphy by Elizabeth Crowley
Set in Minion
Printed and bound by Thomson-Shore
The paper in this book meets the guidelines for
permanence and durability of the Committee on
Production Guidelines for Book Longevity of the
Council on Library Resources.

Most University of Georgia Press titles are
available from popular e-book vendors.

Printed in the United States of America
19 18 17 16 15   C 5 4 3 2 1

Library of Congress Control Number: 2015933069
ISBN: 978-0-8203-4767-7 (hardcover: alk. paper)
ISBN: 978-0-8203-4768-4 (e-book)

British Library Cataloging-in-Publication Data available

This book is dedicated to the Georgia zoologist and natural historian who first introduced me to the Georgia coast: Dr. William D. Burbanck, a professor at Emory University when I was an undergraduate student. It is also dedicated to my husband, Barry, and my sons, Aaron and Jared, who helped me appreciate the wonders of the estuaries, salt marshes, and beaches of Georgia. And I owe very much to all the researchers who spent years and careers studying salt marsh and estuarine flora and fauna along the coasts of Georgia and South Carolina.

CONTENTS

PREFACE. Island Bound ix

1. Marine Habitats of the Georgia Coast 1
2. What You Don't See: Microscopic Life 10
3. Marsh Grass, Live Oaks, Sea Oats 32
4. Creatures of the Black Goo 49
5. Mud Dwellers of Marshes and Creeks 58
6. Creepy Crawlies: Insects and Spiders 73
7. Marsh Life: Scales 79
8. Marsh Life: Feathers and Fur 86
9. What Lies Beneath: Zooplankton 101
10. Attachment to Place: Settlers 109
11. Sound Swimmers: Nekton 115
12. On, and under, the Beach: Living in Sand 132
13. Loggerheads 146
14. Shorebirds 150
15. Seasons in the Sun 157
16. The Once and Future Coast 171

Acknowledgments 191

APPENDIX 1. Where to Go to Enjoy Georgia Coastal Wildlife 193

APPENDIX 2. Conservation Organizations Working to Protect the Georgia Coast 197

APPENDIX 3. Methods for Collecting and Inspecting Coastal Biota 199

Bibliography 203

Index 219

Aerial view of Sapelo Island's Marsh Landing Dock on the Duplin River, August 1978.

PREFACE   Island Bound

In the spring of 1974, as a newly minted PhD, I landed a temporary position as a postdoctoral scholar working on a microbiology project for a professor at the University of Georgia. The job wasn't on the UGA campus in Athens. I was to do research at a coastal laboratory, the University of Georgia Marine Institute on Sapelo Island. In 1968, as an undergraduate student at Emory University in Atlanta, I had glimpsed Sapelo. A zoology professor, Dr. William Burbanck, took his class in aquatic ecology to the Georgia coast for a field trip. Aboard Emory's little motor launch, *Driftwood*, we puttered up the Intracoastal Waterway along the tidal sounds and rivers, past green expanses of salt marsh, to Sapelo's Marsh Landing Dock on the Duplin River (really just a long tidal embayment) to pick up a local guide. I gazed at the island's dark mass of live oak forests lining the marsh edge and wondered what was behind them. I never imagined that in just a few years I would find out.

The Marine Institute has a venerable history, which I was only dimly aware of at the time. The laboratory was set up in 1953 as a collaboration between a research foundation established by R. J. Reynolds Jr. and a University of Georgia ecologist, Eugene Odum. Eugene Odum and his brother, Howard, were pioneers of ecosystem ecology, a field devoted to studying how all the microbes, plants, and animals in a community interact with the local geology and climate to form a stable, self-perpetuating system. Eugene Odum is often called the "father of ecosystem ecology" because he wrote the first and most influential textbook in this field. I studied from a later edition of Odum's book in a biology course at Emory.

In 1934, at the height of the Great Depression, Reynolds bought Sapelo Island from an automobile czar, Howard Coffin. Reynolds updated and enlarged the antebellum mansion on the south end, known locally as the "Big House," as a vacation home. He renovated a cluster of buildings near his mansion that had housed Coffin's guests and a dairy operation. The buildings were arranged around a large square, in the middle of

Main laboratory building of the University of Georgia Marine Institute on Sapelo Island, June 1974. The ground floor of the building was originally used as a barn by Howard Coffin and then by R.J. Reynolds Jr. The Turkey Fountain was installed by Howard Coffin for his hunting buddies. Dogs could lap from the lowest level, two side troughs were for horses, and water for hunters spouted from gargoyle heads just below the crowning turkey at the top. In the 1970s, water was pumped up to the fountain, since by then the natural artesian water pressure was insufficient for the fountain to flow.

which was a three-tiered artesian fountain topped by a cement statue of a wild turkey. The Turkey Fountain had provided water for Coffin and his guests upon their return from hunts on the north end of the island, where wild turkeys abounded. Their thirsty dogs could lap from the bottom level of the fountain, the horses could drink from two long troughs on either side, adorned at the ends with statues of rather mean looking male turkeys, and men could sate themselves with water spouting from the mouths of equally frightful gargoyles into an upper pool around the base of the crowning turkey.

On three sides of the square around the fountain were guest apartments and offices for people working for Reynolds. On the fourth side was a large two-story building. The bottom floor of this structure housed stalls for dairy cows and a milk-processing room. The top floor of one section of the building included a large theater with a screen and a movie projector, to provide evening entertainment for Reynolds and his friends. A couple of seats in the theater were extra wide in order to accommodate his real "fat cat" guests. When the scientific institute was established, the barn was transformed into the main laboratory building. The theater was used as a seminar room for guest speakers and class lectures. Before the age of laser pointers, presenters would employ either a fishing rod or a long bamboo pole to indicate features on their slides. Occasionally, the old movie projector from Reynolds's time would be used on Saturday nights to show the island residents a film. The guest apartments were available for research scientists living on or visiting the island.

A mean-looking wild turkey sits above the mustached gargoyle heads on the Turkey Fountain. The main laboratory is behind. By 1982, the fountain pump was no longer functioning and the basins were dry.

In the early 1950s, one of the first scientists to work at the new science laboratory was Lawrence (Larry) Pomeroy, a marine biologist who had recently earned a doctorate from Rutgers University, in New Jersey. His colleagues were appalled that he would even consider a position that required him to reside on an isolated island on the coast of a southern state. As a boy, Pomeroy had gone on vacations with his parents to Florida. He recalled that on their drives down the East Coast, they would stop at a place in Georgia to enjoy ice cream made with milk from the Reynolds dairy on Sapelo. So he had a fond memory about the island.

Pomeroy and his wife, Janet, arrived in the fall of 1954 at the newly established institute. In the main laboratory building, rats scampered in cow stalls still filled with hay. When Reynolds was in residence on the island, the scientists working at the Marine Institute would be invited up to the Big House on Friday evening for dinner. Pomeroy remembered lavish meals served with fine French wines. Reynolds was so taken with Janet Pomeroy that he named the ferry used to transfer people and goods between Meridian Dock on the mainland and Marsh Landing Dock on the island after her. When I first came

Island Bound   xi

to Sapelo, the *Janet* was the boat that residents routinely took to the island and that delivered our daily groceries and mail.

Another scientist, John Teal, came to the island with his wife, Mildred, in 1955; they stayed for four years. Working in a marsh near the Marine Institute, Teal carried out a landmark study of the ebb and flow of plant carbon in southeastern salt marsh estuaries. A wooden walkway over this marsh still stands; it is called "Teal's boardwalk." The Teals were smitten with Sapelo Island and the surrounding salt marsh estuary. They wrote *Portrait of an Island*, a book describing the natural history of Sapelo and their experiences while living there. After returning to the Woods Hole Oceanographic Institution in Massachusetts, John Teal worked on New England salt marshes. He and Mildred later published a second book: *Life and Death of the Salt Marsh*, about coastal habitats along the Atlantic seaboard and how humans were affecting these ecosystems.

A year after I arrived at Sapelo, I was offered a position as a research scientist on the faculty of the Marine Institute. The pay, $16,000 a year, was better than my postdoc salary, so I grabbed it. No doubt having a woman on the staff in the midst of the feminist movement of the 1970s was an advantage for the laboratory. I was assigned an apartment on one side of the central quadrangle. I became involved in a big research project that Larry Pomeroy, who by then had moved to the main UGA campus, in Athens, together with a colleague, Dick Wiegert, had going in the marshes around Sapelo Island. Their idea was to follow up on John Teal's earlier work in the salt marsh estuary in order to pin down the way this system operated. Teal's research had suggested that most of the plant matter produced in the expansive cordgrass marsh was washed out to the tidal creeks and rivers as dead leaves and stems. The water in the estuary was too murky for much algal growth. Thus, the food web that supported all the crabs, shrimp, fish, and birds that lived in the estuary ultimately had to be fueled by cordgrass production. But data from earlier studies by Teal and other researchers was too spotty to verify this. Pomeroy, Wiegert, and the rest of the participants in the project intended to carry out definitive research to confirm Teal's concept about the cordgrass-detritus food web.

My part was fairly small: $4,000 out of the National Science Foundation grant that funded the project, to test whether the sources of plant carbon in the estuary could be traced by using the ratios of two stable isotopes of carbon: C-12 and C-13. I had gotten the bug about this rather new approach during my first year on Sapelo, when Pat Parker, a geochemist from the University of Texas, gave a talk in the theater of the Marine Institute (I think he used the fishing rod as a pointer) on the use of stable carbon isotope analysis to look at what kinds of plants were eaten by what kinds of animals in a Gulf of Mexico coastal bay. The data I later obtained on ratios of C-12 and C-13 in marsh plants and animals, along with other results from Pomeroy and Wiegert's overall project, showed that the marsh food webs were much more complex than Teal's earlier work had indicated, and were only partly based on cordgrass production.

Soon after, I began working on a salt marsh project with a graduate student of Larry Pomeroy's, Barry Sherr. Island romance ensued, and we fell in love and were married. In 1979, Barry and I spent a year and a half doing research at a laboratory on the Sea of Galilee, known in Israel as Lake Kinneret, at the invitation of an Israeli ecologist, Tom Berman, who was a friend of Pomeroy's. We came back to Sapelo in the spring of 1981, bursting with new ideas for work on the estuary's tiniest plankton, the minuscule flagellated protists that eat the bacteria that swarm in tidal creek waters. Rather quickly, we acquired a border collie puppy, a small sailboat, and two baby boys, both born in Savannah. We did our research, played with our dog, sailed our boat, and raised our kids, all the while soaking in the wild wonders of the Georgia coast.

In 1990 we reluctantly left for faculty positions at Oregon State University. But we cherish our memories of the exuberant wilderness around Sapelo Island. Just as John and Mildred Teal felt an intimate connection with this special place, so do we. This book is meant to give others an understanding of the fascinating life of the region, from the smallest creatures in marsh mud and estuarine water to the mummichogs and multitudes of other animals that find food and shelter in the vast expanses of marsh grass, in the sounds, and along the beaches of the Georgia Isles.

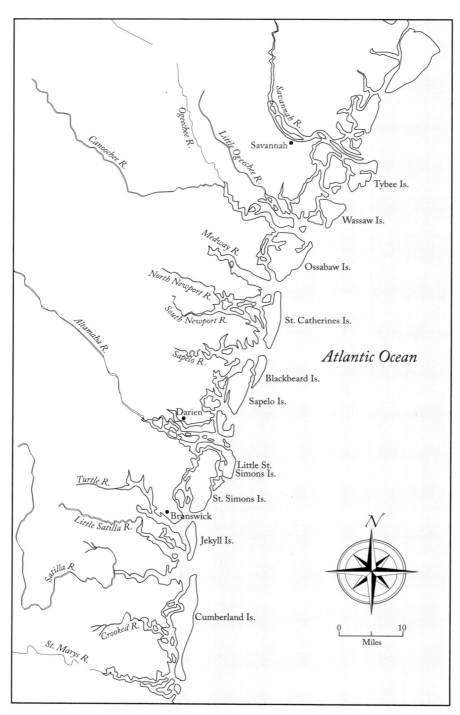

The sea islands of the Georgia coast.

Marsh Mud and Mummichogs

Gulls flying over typical low-energy surf at Nannygoat Beach on Sapelo Island.

1

Marine Habitats of the Georgia Coast

Along the Atlantic coast of the United States, from Cape Hatteras, North Carolina, to Miami, Florida, is a stretch of the continental shelf known to oceanographers as the South Atlantic Bight. The central part of this region, from Cape Romain, South Carolina, to Jacksonville, Florida, just over the Georgia state line, curves deeply westward, well away from the Gulf Stream; here the shelf is broad and shallow. The shape of the coastline and shelf in this area has two important consequences. First, the twice-a-day tidal surge traveling over the oceans, pulled by the gravity of the moon and sun, is reflected downward from Cape Romain and builds up over the long slope of the shelf. When this concentrated bulge of seawater reaches the Georgia coast, the tidal rise along the shore is an impressive six to nine feet. At the same time, the shallowness of the near-shore continental shelf saps the energy from large waves traveling from the open ocean, so by the time they break on sea island beaches, they are tamed into small wavelets lapping over the sand. No surfer's paradise here but rather a warm gentle ocean for wading and swimming.

Over time, the tide and winds created sand dunes on which plants grew. More sand accumulated, eventually building islands at the edge of the sea. Behind the line of oceanfront islands lay a sheltered expanse of muddy salt marsh and open estuary several miles broad. This type of coastline has existed along the edge of the Georgia Bight for a long time. As the sea's height has fallen or risen over the past million years, the line of islands and band of marshes have advanced or retreated. The sea islands that exist today are actually two lines of islands: the old main islands and the newer beachfront dunes and sandbars. Old Pleistocene dunes form the heart of the main-island interiors. The fringing dunes and beaches were established during the modern, or Holocene, period within the past four thousand to five thousand years. Past shorelines can be discerned farther inland on the Georgia coastal plain as ridges of sand dunes that were island cores when sea level was higher. The salt marshes between the islands and the mainland grow on deposits of sand and silt eroded from the Appalachian Mountains and the red clay soils of the Georgia foothills, brought down to the coast by rivers.

The long reach of the tides nourishes the plants and animals that live in Georgia estuaries. Twice-daily tidal flooding maintains the vast area of salt marshes behind the sea islands. Saltwater fills and drains the marsh via a network of creeks and long bays, or tidal rivers, created by the strong sea currents. At the same time, the tidal flows scour the marsh creek banks and bottom of the estuary, shifting mud and sand sediments about and keeping lighter particles in suspension. Because the bottom substrate is unstable, and the water clouded with plant detritus and silt, there are few attached, or sessile, plants and animals along this coast as compared with regions farther north marked by clear water and rocky shores or with transparent coastal oceans farther south that feature hard sand bottoms and coral reefs.

The subtropical climate of coastal Georgia has a major role in determining which species live here. Mild winters with few frosts allow palms to grow along the edges of the marsh. Ducks and shorebirds migrate south in the fall to overwinter in Georgia estuaries and beaches. But the blazing summer sun heats both air and water, limiting the presence of northern species that need cooler temperatures to survive. Seashore animals of cold New England waters generally don't appear south of Cape Hatteras.

Thus, three distinctive types of marine habitat characterize the Georgia coastline—intertidal salt marsh, subtidal estuary, and open beach and dune—and each of these areas includes distinctive kinds of life. In addition, scattered among the saltwater species are such inland creatures as herons, alligators, raccoons, and deer, which have adapted to these coastal environments.

Intertidal Salt Marsh

The great flats of salt marshes, green in summer, golden in fall, are an impressive feature of the Georgia coast. Two-thirds to three-fourths of the estuary lying behind the sea islands is salt marsh, and over 90 percent of the marsh is covered by just one species of plant: smooth cordgrass, *Spartina alterniflora*. Its genus name, *Spartina*, derives from a similar salt grass growing along the shores of the Mediterranean Sea. The ancient Greeks found that fibers from the dried grass could be woven into a strong rope. *Spartine* is Greek for "cord," so both the scientific and common names of the plant refer to this practical use.

Although cordgrass can't survive completely immersed, it is superbly adapted to intermittently flooded, waterlogged, silty, salty soils. *Spartina* grows rapidly by sending out thick rhizomes wherever it can establish itself, from the height of mean low tide to height of mean

The author sampling in a cordgrass marsh plain near Marsh Landing on Sapelo Island, fall 1976.

Marine Habitats of the Georgia Coast

high tide. The salt marsh is divided into elevation zones. Cordgrass plants lining tidal creeks are tall and bright green. Just behind this luxurious growth is the levee marsh, a hump of mud. The hump forms where high tide drops its load of suspended sediment, a result of the incoming water slowing as it moves through the thick stands of creekside *Spartina*. Levee marsh grass is not quite as lush. Farther back is the broad stretch of marsh where *Spartina* plants grow thickly but are much shorter in height. This "marsh plain" is also called "low marsh" to distinguish it from the higher zones near the marsh border, but the term "low marsh" can be confused with the very low creekside marsh. Here I use "marsh plain" to refer to the salt marsh beyond the levees. The marsh plain is dominated by smooth cordgrass but can include other plants, such as black needlerush.

The winding of small tidal creeks into the marsh plain is marked by lines of taller, greener cordgrass along the channels. Plants that are more frequently flooded have greater access to nutrients brought by the tides. Daily surges of seawater into the marsh perform another service: washing away excess salt. *Spartina* plants growing in marine mud need to constantly rid themselves of salt taken up by their roots from the marsh soils. They do this by excreting salt from special glands in their leaves. For creekside cordgrass, salt accumulated on the edges of the leaves is washed off by the flow of estuarine water twice a day. Plants growing in the marsh plain are covered less often by the tide and must endure salt crystals on their leaves until the next high-flood water arrives.

At the landward edge of the marsh, the soil is inundated very infrequently by the highest tides. High summer temperatures evaporate water from the exposed upper-marsh surface, leaving seawater salts behind. The soil becomes too salty for *Spartina*. Plants able to tolerate high soil saltiness become established in small patches. Common upper-marsh plants are glassworts, saltwort, salt grasses, and sea oxeye aster. Where the soil is too saline for even these salt-tolerant plants, white sandy zones barren of vegetation mark the marsh edge.

At the landward edge of the marsh thick stands of black needlerush, a plant that doesn't tolerate regular inundation by seawater, often

Aerial view of *Spartina* salt marsh on the south end of Sapelo Island, August 1978. Note the tall dark creekside cordgrass and the shorter, lighter-colored meadow cordgrass. Dark patches near the head of the creek are stands of black needlerush.

replace cordgrass. Finally, bordering the marsh one finds land plants that are flooded only at the highest spring tides and that can tolerate some salt. Two tall composite plants in the daisy family—groundsel bush and marsh elder—along with wax myrtle bushes, are found along the marsh edge. Here and there small islands of high ground in the marsh, called hammocks, offer refuge for shrubby junipers, palmettos, and even small live oak trees.

The most conspicuous animals of the salt marsh are great blue herons and white egrets, which hunt small fish in the tidal creeks at low tide and in the marsh at high tide and the multitude of comical fiddler crabs that emerge at low tide to forage on the marsh mud. There are a great number of other inhabitants and visitors, including the raucous but elusive clapper rail and the occasional raccoon snacking on crabs among the cordgrass stems. Most of the animals of the Georgia coast depend on the salt marsh, in one way or another, for food or shelter.

Subtidal Estuary

At low tide, about one-third to one-fourth of Georgia estuaries are open water. The murky, greenish-brown water of the subtidal estuary teems with hidden life. Microscopic algae and zooplankton support a food web that produces shrimp, crabs, and fish. Oyster reefs damming tidal creeks, along with clams in sandy flats along creek bottoms, filter out the abundant tiny organisms from the water. Seagulls, terns, and pelicans fish in the estuary. Bottlenose dolphins play in the open sounds, and alligators glide slowly along tidal rivers. Dense schools of young menhaden, a fish that lives on plankton and is prized for its oily flesh, darken the water surface. In the spring, thousands of horseshoe crabs congregate in the shallow waters of tidal rivers and along ocean beaches to spawn; very large horseshoe crabs can often be found on the shore.

Plants and animals that grow by attaching to surfaces have a difficult time in these estuaries because of a lack of suitable substrate. Large algae, sponges, corals, bryozoans, and hydroids occur only on rare hard surfaces, such as oyster reefs and dock pilings. The number of benthic-dwelling species is fairly small because of the paucity of types of bottom habitat. Only on occasional rocky outcrops miles offshore, such as Gray's Reef National Marine Sanctuary off Sapelo Island, can one find a rich diversity of benthic animals.

The Ocean Shore

The beach and dune habitat of the sea islands is only a small part of the coastal ecosystem, but it is the one most familiar to people. The beach has two zones: the wet, intertidal beach slope and the dry, sandy berm between the high tide line and the first dunes. The lower beach is dynamic, often featuring shallow pools or long sloughs running parallel to the surf line and drained by deep runnels out into the sand. At low tide, the holes and tubes of burrowing bristle worms and ghost shrimp dot the beach. These sedentary animals live on microscopic life in the low, breaking waves and in the sand. In late summer, benthic algae attain such high densities that the lower beach becomes stained green or brown with their pigments. Flocks of sandpipers run after the retreating surf on the lower beach, probing the wet sand for tiny clams and crustaceans.

When the tide goes out, lines of sand bars appear above the waves, exposing subtidal habitats. Many species of fish, mollusks, crabs, and starfish live in the ocean just off the beach; their bodies, shells, and egg cases sometimes wash up on shore. To avoid the disruptive currents of tides and breaking waves, some of the near-shore animals hide in the sand, burrowing through sediments or living in burrows or tubes from which they emerge to capture food from the restless water above.

On the upper beach, dead cordgrass stems and other debris accumulate as piles of wrack deposited by high tides, providing food and shelter for beach crabs, amphipods and insects. The wrack also provides favorable conditions for the germination of sea oat seeds, thus fostering new plant-anchored dunes. All along the high beach, deep holes, from a half-inch to two inches across, mark the burrows of ghost crabs. These boxy, light-colored crabs are related to the fiddler crabs of the cordgrass marsh. Ghost crabs usually emerge only at night to hunt small animals in the surf or to scavenge among the beach wrack.

On calm summer nights, monsters arrive from the sea. Enormous loggerhead sea turtles laboriously shove themselves across the beach to the base of the dunes. There, each female painstakingly scoops out a deep depression with her hind flippers and deposits dozens of eggs. If the nest is not disturbed by ghost crabs, raccoons, or humans, the baby turtles will hatch and scamper into the sea in late summer. Isolated barrier island beaches are also good nesting sites for shorebirds: gulls, terns, skimmers, and plovers.

At the landward edge of the beach, sand dune habitats include the unstable fore dunes, the more established back dunes, and moist interdune meadows. Sand dunes are rather inhospitable for plants: hot and dry, exposed to wind-whipped salt spray, and poor in mineral nutrients needed by plants to grow. Despite this, several species are successful colonizers; their stems and roots maintain the sand dunes against the eroding action of wind and storm waves. Most common are the graceful sea oats. Sand dunes also harbor animals, such as the ubiquitous mole, which often burrows well out onto the beach in search of tasty tidbits. Mice, cotton rats, and marsh rabbits eat the seeds and leaves of dune plants, and in turn are preyed on by hawks and snakes. The back dunes provide excellent hunting grounds for large rattlesnakes.

The following chapters cover the most common types of life in the marine habitats of the Georgia coast, from tiny microscopic organisms to large vertebrates. Over decades, scientists at the University of Georgia Marine Institute on Sapelo Island, as well as at other institutions in the Southeast, have uncovered the secrets of many of these creatures. Scientific as well as common names are given for the species, since common names can vary. Taxonomic names are not immutable either, since new information often changes how scientists relate species to one another. One example: the ribbed mussel, abundant in cordgrass marshes, was initially named *Modiolus demissa*, which placed this mollusk in the same group as other common marine mussels. Later, taxonomists decided that the ribbed mussel should be in its own genus, and it was renamed *Geukensia demissa*. Even with such changes, scientific names are essential for determining the precise species present in an environment and for understanding their relation to other, closely related species.

We will start our tour of coastal marine life with microbes, the unseen, usually ignored, but central players in the ecosystems of these (and all other) habitats.

2

What You Don't See *Microscopic Life*

If you scoop water from beach surf or from a tidal creek into a clear glass bottle and hold it up to the light, chances are you won't see any animals flitting about. The water may look empty of life, but in fact it is teaming with minute organisms. In one teaspoonful of water there may be thousands of microscopic algae and colorless protists, a million bacteria, and ten million viruses. Marine microbiologists often familiarly refer to all these little guys as "seabugs." They are the smallest, but by no means the least significant, forms of life in any ecosystem, salt marsh estuaries included. Some of the relatives of these simple organisms get large enough to be noticed; also included in this discussion are fungi growing on dead cordgrass and in beach wrack, and large algae that colonize cordgrass stems and docks during the winter.

Oceanographers have long known that microscopic algae are abundant in the ocean and form the basis of virtually all marine food webs. It was a surprise when, in the 1960s, tiny bacteria were found to be even more numerous than algae in seawater. The vital role of marine bacteria in plankton was shown, in part, by research done at the Marine Institute on Sapelo Island.

In the years after World War II, oceanography thrived, along with other natural sciences, and research in the emerging field of marine microbial ecology expanded. The ecosystem concept—how ecosystems were structured and how they functioned—was a crucial step in the progress of understanding the roles of microbes in the sea. The textbook *Fundamentals of Ecology* (1953) by Eugene and Howard Odum excited interest in systems ecology. The newly established University of Georgia Marine Institute, funded by R. J. Reynolds's

Sapelo Island Research Foundation, offered a perfect place for scientists to study ecosystems.

One of the earliest projects at the laboratory on Sapelo was John Teal's pioneering work on energy flow through the salt marsh ecosystem. In a 1962 paper in the journal *Ecology*, Teal reported that very little of the plant matter produced by cordgrass each year was directly eaten by insects or other animals. Instead, the twice-daily tides flooding the marsh carried off dead cordgrass to estuarine creeks and rivers. In his paper, Teal included results of experiments showing that *Spartina* leaf fragments were decomposed by bacteria. In the process, living and dead microbes enriched the fragments with protein, making the plant material more nutritious than nondecomposed cordgrass for estuarine animals. Later work by Eugene Odum and a PhD student, Armando de la Cruz, confirmed that much of the production of *Spartina* was indeed carried out of the marsh and was decomposed by microbes in the estuary. The consensus derived from this and similar research carried out at the Marine Institute was that the food webs of salt marsh estuaries were largely based on this microbially enriched plant detritus.

In 1954, as a young PhD, Lawrence (Larry) Pomeroy was offered a position in the initial group of researchers at the institute. Pomeroy was soon joined by a postdoctoral colleague, Robert (Bob) Johannes. Both Pomeroy and Johannes were interested in elemental cycles in aquatic ecosystems. Johannes carried out experiments with bacteria-eating flagellates cultured from estuarine water. His data showed that the flagellates were capable of excreting phosphate, a plant nutrient, at very high rates. The rate of phosphorus release per unit of flagellate mass was found to be much greater than similar biomass-based rates of phosphate excretion determined by other scientists for animals in the zooplankton. The result was dramatic evidence that the smallest organisms in the plankton—notably, single-celled microbes—had a much higher metabolic activity, or "rate of living," than did larger plankton.

This finding led Pomeroy and Johannes to collaborate on experiments demonstrating that the respiration rates scientists routinely measured in seawater were mainly due to the activity of single-celled microbes rather than that of planktonic animals. Respiration, or "breathing," which is carried out by all organisms in oxygen-rich environments, can be quantified by the rate of disappearance of oxygen over time. Pomeroy and Johannes's experiments consisted of filling a series of glass bottles with seawater and stoppering them so that they were airtight, thereby keeping oxygen from leaking in or out. The bottles were incubated in the dark to prevent any algae in the water from producing oxygen by photosynthesis. Then, the total amount of oxygen in individual bottles was measured using a chemical assay. The bottles were sampled, one by one, every few hours over a set time period, usually twelve to twenty-four hours, until all the bottles had been assayed. The results were plotted on a graph of oxygen content in each bottle versus the elapsed time since the bottle was sampled. Typically, they found a linear decrease in oxygen content in the bottles over time because of respiration by the organisms. The decrease in oxygen was then used to calculate the rate of respiration in the seawater.

In their experiments, Pomeroy and Johannes compared rates of oxygen decrease in unfiltered seawater, which contained all sizes of organisms, with rates of decrease in seawater passed through a no. 2 plankton net with a mesh size of 366 microns (a micron is one millionth of a meter), a bit larger than the size of the mesh in a nylon stocking. The netting prevented any animals larger than that size, which included most zooplankton, from going into the sample bottles. The researchers discovered that the oxygen decrease, that is, the respiration, in the bottles with the net-screened water was nearly identical to the respiration in unfiltered seawater. From this result, it was clear that the organisms that passed through the plankton net were responsible for consuming virtually all the oxygen during the incubations. Pomeroy and Johannes concluded that, by volume, the smaller organisms in the plankton, mainly bacteria and their flagellate predators, had a respiration rate ten times greater than that of the excluded zooplankton.

By the early 1970s, Larry Pomeroy had sensed that the prevailing concept of marine food webs was not accurate. That established idea of food webs in the ocean was formalized in a slim 1974 book: *The Structure of Marine Ecosystems*, by John Steele, a noted Scottish mathematical modeler interested in marine fisheries. Steele's diagram of food webs relegated microbes to a "bacteria" component of the benthos (bottom-dwelling organisms) responsible for decomposing fecal material. In his model, bacteria and other heterotrophic microbes, like bacteria-eating flagellates, played no formal role in the plankton. Steele concluded: "The phytoplankton of the open sea is eaten nearly as fast as it is produced, so that effectively all plant production goes through the herbivores." By "herbivores," Steele meant crustacean zooplankton that feed on algae.

Pomeroy knew that Steele's model of marine food webs was incomplete. There was a lot of dead organic matter drifting about in seawater. Based on his and Johannes's experiments, Pomeroy had evidence that decomposition of this organic matter by microbes caused most of the respiration in seawater. Pomeroy then had a profound insight. He combined the ecosystem theory promoted by Eugene and Howard Odum with the detritus-based food web concept developed by John Teal to form a new view of the role of microbes in the sea. Pomeroy extended Teal's detrital food web of the salt marsh estuary to the ocean in general. In the revised model, bacteria decomposed dead organic matter originating from phytoplankton. Then bacteria and their protist predators reprocessed most of the nitrogen and phosphorus in the organic detritus back to chemical forms easily used by phytoplankton for further growth. Pomeroy's idea directly challenged the model of marine food webs depicted in John Steele's book.

Around this time, John Bardach, the editor of the journal *BioScience*, was looking to increase his readership by publishing review papers with wide appeal. (A review paper summarizes the current state of knowledge in a field.) He asked his friend, Bob Johannes, to suggest potential authors of such reviews. Johannes replied that his colleague Larry Pomeroy had new ideas about marine ecosystems that might be of interest. Pomeroy obliged the editor with a manuscript summarizing evidence supporting his concept that bacteria, along with their flagellate predators, were central to the functioning of marine ecosystems. Pomeroy later told me that his manuscript was sent out to two external referees for their opinion. One of the reviewers never responded, and the other one said the paper was nonsense and should be rejected. But Bardach, thinking that a controversial review was just the thing to excite interest in his journal, published Pomeroy's submission anyway, referees be damned.

By the end of the 1980s, Pomeroy's 1974 paper in *BioScience*, "The Ocean's Food Web: A Changing Paradigm," was regarded as a pivotal description of new ideas about how most organic matter in planktonic food webs was processed by small planktonic organisms, mainly by marine bacteria and their protist grazers. Research on marine microbes was greatly stimulated by Pomeroy's new concepts, which directly resulted from scientific investigations carried out in Georgia marshes and estuaries.

The story of marine viruses is even more recent, dating from only the late 1980s. While on the faculty of the University of Georgia Marine Institute, I played a small role in that development.

An eye-opening paper appeared in the journal *Nature* in the summer of 1989. The authors, a group of Scandinavian scientists, reported that viruses were incredibly abundant in seawater. Most viruses are small bits of DNA encased in a hard capsule (although some DNA viruses lack capsules and some types of viruses, such as HIV, the cause of AIDS, are based on RNA instead of DNA). The viral DNA can enter cells, hijack the cell's metabolic machinery, and cause the cell to produce many more copies of the virus. The cell eventually fills up with viruses and bursts, releasing innumerable viruses to seek out and infect healthy cells. The editors of *Nature* like to showcase newsworthy articles by inviting an expert to write a comment on the importance of the finding. In this case, they asked me to write about the virus paper. I wrote up a summary titled "Aquatic Viruses: And Now, Small Is Plentiful," which came out in the same issue as the virus article. Soon afterward my office phone at the Marine Institute was ringing. People are well aware that viruses are the cause of common human diseases such as colds and the flu. The idea that playing in the surf at the beach meant exposure to gazillions of viruses was pretty upsetting to the public.

One of the calls was from the program *CBS News Nightwatch*, which was produced in Washington, DC, and hosted by Charlie Rose. *Nightwatch* asked me to do an interview about viruses in the sea. So I did. The show was taped in D.C. on a hot summer afternoon. I was asked what kind of makeup I wanted for the cameras. I didn't know. So they decided I should have "nun's makeup," the minimum necessary, which was fine with me. Mr. Rose was really nice while I carefully explained that no, the marine viruses the Scandinavians discovered could not infect humans. The only organisms that had to worry about these viruses were marine bacteria. Almost all the viruses in seawater are bacteriophages, which enter and kill bacterial cells. Beachgoers had no cause for concern.

Bacteria

Viruses are not really alive; they only have the potential for replication inside other cells. The simplest living cells are the bacteria. The scientific term for bacteria is prokaryotes, that is, cells without a nucleus (or *karyon*, a Greek work meaning "seed") or internal structures (organelles). These microbes are single cells so small that four million of them could easily fit on the head of a pin. In the 1980s, molecular geneticists were surprised to discover that there were not one but two distinct groups of prokaryotes: bacteria and archaea. Each group was given its own domain, "domain" here indicating a level above kingdom in the familiar classification system of kingdom, phylum, class, order, family, genus, and species. Besides profound differences in their genetic codes, these two groups can occupy different ecological niches. Whereas bacteria are found all over the biosphere, many groups of archaea are specialists, growing in environments unusable by other organisms: the depths of the ocean, oxygen-deficient habitats, or conditions of extremely high temperature, such as hot springs.

Species of bacteria and archaea carry out their life functions in an amazing variety of ways. Although tiny and structurally simple, prokaryotes have the most varied and exotic biochemistry of any lifeforms on this planet. Most prokaryotes feed on nonliving organic matter, but others make their cellular material from carbon dioxide and inorganic nutrients by using solar energy, as *Spartina* and other higher plants do. Some types of bacteria and archaea can grow without sunlight, using energy generated by reactions that involve inorganic chemicals. Exobiologists suspect that if there is life on any of the other planets or moons in our solar system, it is probably not little green men but rather tiny single-celled microbes with metabolic capabilities similar to those of earthly bacteria or archaea.

In all salt marsh estuaries, prokaryotic microbes, along with eukaryotic fungi, degrade much of the biomass of marsh grass, which grows during summer and then dies back in the fall. Most estuarine and marsh animals can't digest the tough marsh grass, but many animals can live on the bacteria, fungi, and other microbes growing on the decomposing plant fragments. These detrital grass particles and their attached microbes are partly responsible for the brown color of the water in Georgia estuaries.

Bacteria also live deep in the marsh soils, where there is little or no oxygen. These types of bacteria can't "breathe," or respire, like microbes and animals bathed in oxygen-rich air or water. They have other ways of getting on. Many of the mud-dwelling bacteria are fermenters, akin to the bacteria and yeasts that make alcohol and acids in the process of producing beer, wine, and vinegar. The marsh fermenters down in the sediment convert simple sugars to fatty acids and alcohols. But the marsh soils don't become alcoholic.

The alcohols and other simple compounds produced by the fermenters are quickly used up by another group of bacteria, the sulfate reducers. Sulfate-reducing bacteria are highly specialized. They can combine an oxidized sulfur compound, sulfate, which is abundant in the seawater permeating the marsh mud, with by-products of fermentation—organic acids and alcohols—to gain energy for growth. In doing this, the sulfate-reducing bacteria convert the oxidized sulfate to a reduced sulfur compound, hydrogen sulfide, which is poisonous to most life. So instead of alcohol building up in the belowground marsh, toxic sulfide accumulates. If you dig into the surface of a *Spartina* marsh or a tidal-creek mud bank, the surface quarter inch will be brown, but below that the mud is black and stinky. The smell is like that of rotten eggs, and for the same reason: sulfide. The black color comes from reduced iron in the silty, iron-rich clay. Fortunately, much of the sulfide becomes bound up in an iron-sulfur compound, pyrite. But there is still enough free sulfide in the mud to cause a powerful smell.

The sulfate-reducing bacteria have another impact. Salt marshes produce very little swamp gas, methane, which comes from the activity of a curious group of prokaryotes in the archaea. These microbes are profoundly metabolically challenged. Methane producers use only very simple biochemistry, and only in habitats where there is no oxygen. They combine free hydrogen ions with carbon dioxide, yielding methane, which they excrete, and gaining barely enough energy for growth. But in oxygen-depleted sediments bathed in sulfate-rich seawater, the methanogens are out of luck. Sulfate-reducing bacteria use a great deal of hydrogen when reducing sulfate to hydrogen sulfide. The sulfate reducers grow aggressively, sopping up all the free hydrogen in the anoxic (oxygen-free) marsh soils, depriving the poor methane-producing archaea of food resources. In marshes and swamps up coastal rivers, where the water is fresh and contains little sulfate, methanogens can go about their business in the mucky mud, undeterred by the gluttonous sulfate-reducing bacteria. Unlike tidal salt marshes, river swamps exhale large quantities of methane.

One usually can see bacteria only with a microscope. But certain species of bacteria in the estuary form colonies of millions of aggregated cells, often growing in long filaments. Whitish streaks and clumps on the surface muds of marsh creeks are colonies of sulfide-oxidizing bacteria. These bacteria don't live on plant detritus; instead, they make their own organic matter by using the chemical energy stored in reduced sulfur compounds to fix carbon dioxide. This process, a form of chemosynthesis, captures energy from the reaction of sulfide oozing up from the marsh soils below with oxygen from the air and water above. The sulfide originates, of course, from the activity of those sulfate-reducing, sulfide-producing bacteria down in the mud.

The sulfide-oxidizing bacteria on the marsh surface combine sulfide with oxygen and then use the energy thereby released to fix carbon dioxide into organic molecules. This process is akin to photosynthesis, by which plants use light energy from the sun to fix carbon dioxide into sugars. The end product of the reaction of sulfide and oxygen, the oxidized sulfur compound sulfate, is returned back to the seawater. This is an example of biogeochemical cycling, in which different groups of microbes convert a chemical element, in this case sulfur, back and forth between oxidized and reduced states.

The chemosynthetic sulfide-oxidizing bacteria in the marsh are closely related to the sulfide-oxidizing bacteria that support the spectacular tubeworms, clams, and other animals living around hydrothermal vent systems in the lightless deep ocean. At the vents, the sulfide comes from superheated seawater spewing from volcanic fissures in the sea floor. While surrounding the glowing lava just below the seabed, the boiling seawater becomes loaded with reduced chemicals, such as sulfide. White mats of sulfide-oxidizing bacteria around the fissures avidly take up the sulfide from the chemical-laden water and combine it with oxygen in seawater in order to grow by chemosynthesis. The white color of the bacterial filaments is due to tiny light-refracting spheres of sulfur inside the cells. The bacteria strategically take up sulfide in excess when hot fluids vent from the fissures. The toxic sulfide is converted to nontoxic elemental sulfur inside the bacterial cells. When the vent flow of reduced compounds slows down, the bacteria can continue growing by combining their stored sulfur with oxygen. Other sulfide-oxidizing bacteria have managed to form symbiotic relationships with the forests of red-plumed tubeworms and the colonies of white clams that crowd around the vents. The animals provide shelter for the bacteria, which grow in their tissues, and the bacteria are digested by the animals as their only source of food. If you see white splotches of filamentous bacteria on the marsh mud, you have found close cousins of the microbes that sustain the weird and wonderful hydrothermal-vent fauna in the deep sea.

In winter, patches of dark blue-green dot the surface of the mud in the lower marsh. These are colonies of filamentous cyanobacteria.

These light-harvesting microbes used to be called blue-green algae; however, they are not algae at all, but collectively form a major taxonomic group of bacteria. At least three species of cyanobacteria form distinctive colonies on salt marsh mud: *Anabaena oscillarioides*, *Microcoleus lyngbyaceus*, and *Schizothrix calcicola*. Cyanobacteria also occur as blue-green scum on high-marsh salt flats or on the sides of shallow, semipermanent lagoons in the upper beach.

Cyanobacteria were among the earliest forms of life on earth. Putative fossils that resemble filaments of modern cyanobacteria have been found in rocks over 3 billion years old. These photosynthetic microbes grew all over shallow seas during the Proterozoic (2.5 billion to 541 million years ago), fixing carbon and producing oxygen as a by-product for two billion years, which gradually increased the amount of oxygen in the atmosphere to the point that animals, including us, could live. Most of the oxygen inventory of the modern atmosphere was built up by cyanobacteria in ancient oceans. So these lowly organisms deserve our deepest respect.

Eukaryote Microbes

In addition to the prokaryotes, the bacteria and archaea, the third major group of microbes comprises the unicellular eukaryotes. These include both predatory protists and fungi, which eat living and dead plant matter (known as heterotrophs, Greek for "other-feeding"), and algae, which make their own food by photosynthesis (known as autotrophs, Greek for "self-feeding"). The cells of eukaryotes are larger and more complex than those of prokaryotes. Their DNA, like that of higher plants and animals, is in a membrane-bound nucleus (hence their name, *eu-karyote*, Greek for "true-seeded"). They have organelles, including light-harvesting chloroplasts and energy-generating mitochondria. Eukaryotic microbes can exhibit complex behaviors, responding to environmental signals such as light and chemicals.

Fungi include yeasts, molds, and mushrooms; these primitive eukaryotes evolved along a developmental line of their own. Molds, including mildew, and mushroom-producing fungi are common in terrestrial and freshwater environments but rare in saltwater. Molds and mildew are made up of a thin layer of fungal hyphae (threadlike structures) that coat the surfaces of dead fruit, seeds, and leaves (and the walls of buildings during warm humid summers). Most estuarine fungi are too small to be noticed. One sign of the presence of fungi is the changing color of dead *Spartina* stems in the marsh. At first, the dry stems are a light gray, but turn darker as fungal filaments invade the dead tissue. On the beach, a few species of fungi live on rotting plant debris in the dunes, producing fruiting bodies that poke up through the sand in the fall; these include the mushroom-producing fungi *Psathyrella ammophila* and a species of the genus *Melanoleuca*.

Single-celled eukaryotes that prey on other microbial cells used to be called protozoa, from the Greek for "first animal life." But these microbes are not animals, and their diverse lineages are not closely related to animal phyla. Thus the term "protist" is now commonly used for these microbes. Protists range in size from tiny flagellates only a bit larger than bacteria to cells that are just barely visible as floating motes in seawater. Protists include species that are strictly photosynthetic, the unicellular algae. But many species lack light-harvesting chloroplasts and so must live by preying on other microbes. Three major types of these predatory protists are flagellates, ciliates, and amoebae.

Most flagellates are very small spherical or oblong cells with one or two whiplike flagella used to propel them through the water. Some flagellates, such as the choanoflagellates, are voracious consumers of bacteria. Choanoflagellates have a delicate, membrane-like conical "collar" around the flagellated end of the cell ("choano" is from the Greek word for "funnel"). Colonies of these collared flagellates can be found attached to particles suspended in the water. Sponges, some of the simplest multicellular animals, are thought to have evolved from associations of choanoflagellates.

The dinoflagellates are more closely related to ciliates than to other types of flagellated microbes. These protists are the whirling dervishes of the microbial world. "Dino-" is from a Greek word meaning "rotating"; dinoflagellates were so named because they use their whiplike flagella to swim in a peculiarly looping way. Dinoflagellates have large nuclei containing a great deal of DNA, and are known for group behaviors such as swarming and migrating together up and down in the water. While many dinoflagellates have light-collecting pigments, and so can grow by photosynthesis, other species, having lost this ability, live exclusively by hunting and eating other cells, including their dinoflagellate kin. One of the largest predatory dinoflagellates, easily seen without a microscope, is *Noctiluca scintillans*, whose name means "scintillating night light." This large dinoflagellate can capture planktonic animals as well as other protists. *Noctiluca* is bioluminescent, capable of converting biochemical energy into light, just as fireflies do. The soft blue glow seen in coastal waters on summer nights is caused by this dinoflagellate.

Like algae, species of dinoflagellates with chloroplasts can grow by harvesting light, but most are also able to capture and eat other cells, just like their nonpigmented relatives. These photosynthetic dinoflagellates have it both ways. Some species have eyespots that allow them to sense the direction of light in the water during the day. They use eyespot direction to navigate up toward light for photosynthesis, and then down to darker subsurface layers of water where fertilizing nutrients accumulate. Many larger dinoflagellates are "armored" with rigid plates of a cellulose-like material, similar to the stuff that strengthens the stems of land plants. The cellulose plates cover the outer surface of armored dinoflagellate cells in intricate patterns. Some plate-covered dinoflagellates, such as *Ceratium* species, have long "horns" protruding from their cells.

Dinoflagellates often bloom in irruptive outbreaks known as red tides, named for the unique red-brown pigment, peridinin, in the group's chloroplasts. They swarm together, at times changing the color of tidal creeks from greenish brown to a deep red. Some of these swarming species make a variety of toxins harmful to their predators, to other phytoplankton that may compete with the dinoflagellates for nutrients, and ultimately to humans who eat oysters and clams that fed on the dense mass of red tide cells. So far, such toxic red tides have not occurred along the Georgia coast. Dinoflagellates that bloom in upper tidal creeks in spring are from a nontoxic species in the genus *Kryptoperidinium*.

Ciliates are single-celled eukaryotes that move by beating hundreds of tiny hairs, called cilia, on the surface of the cell. These protists (our sons delighted in calling them "silly-its") come in an amazing variety of shapes and sizes. One group of ciliates, the tintinnids, constructs delicate tube-shaped structures, called lorica, around themselves. The most common ciliates in estuaries are small slipper-shaped cells similar to the familiar laboratory ciliate *Paramecium*. (Often, when you see film clips of unspecified swarming "microbes" in documentaries or movies, the microbes are really just *Paramecium* ciliates; they are easy to grow to high density and big enough to be easily seen under a low-powered microscope for filming.) Another type of ciliate, the peritrichs, attach to surfaces, either singly or in large, multi-stalked colonies. Most ciliates flitting about in the plankton are not like *Paramecium*, but are spherical or conical cells with a crown of stiff cilia at the oral, or feeding, end, and look something like the ball-shaped vacuum cleaner attachment used to suck dust off furniture. Ciliates deploy their cilia both to swim and to create local water currents that bring prey cells close enough to capture.

The vacuum-cleaner-ball type of planktonic ciliate has been the object of much research attention. Many of these ciliates grow well in laboratory flasks and reproduce quickly on a diet of algae. Scientists have used these ciliate cultures to explore general relationships between food consumption and rates of growth in relation to the amount of algal food given to the ciliates. Such studies suggest that these ciliates are the main predators of unicellular algae in the sea, and the much smaller bacterial cells are consumed by tiny flagellates.

The following story is an example of how ideas, such as the concept that ciliates don't feed on bacteria, can be challenged by scientific investigation. After returning to the Marine Institute from our stay in Israel, my husband, Barry, and I carried out a simple project that demonstrated the lack of a sharp division of microbial prey between ciliates (algae) and flagellates (bacteria). We dyed estuarine bacteria so they would light up in protist food vacuoles when they were gobbled up. We added the dyed bacteria to estuarine water, and then inspected captured protists to see which ones had fed on the bacteria. We thought that only the little flagellates would be lit up, but found, to our surprise, that many of the planktonic ciliates, too, had eaten our bacteria, often quite a lot of them. Incidentally, our samples contained many very small ciliates, much smaller than those reported by other researchers, and it was those tiny hairballs that were consuming bacteria.

When we submitted our findings for publication, a highly regarded marine microbial ecologist pooh-poohed the work. In his review of our manuscript, he stated that it was impossible for marine ciliates to be as small as we claimed—it would be as incredible as the existence of a mammal one centimeter (a half inch) long. We wrote back to the journal editor that in fact a newborn pigmy shrew, an insectivorous mammal, is about one centimeter long. That stopped the criticism, and our paper was published. Soon after this exchange, the famous ecologist published a book on aquatic protists. He included a figure comparing the relative sizes of protists and mammals. On the mammal side, the pigmy shrew was listed as the smallest.

Among the largest marine protists are amoebae, cells that move and feed using blob-like extrusions termed pseudopodia, Greek for "false feet." In the open ocean, amoebic relatives, the radiolarians and foraminiferans, make distinctive, often beautifully formed shells (also called tests) from the minerals silica and calcium carbonate. Some of these protists are up to half an inch across. The slow but steady downward drift of shells of radiolarians (silica tests) and foraminiferans (calcium carbonate tests) helps form deep-sea sediments. Small foraminiferans are common in salt marshes. Most amoebae in the estuary are simple naked forms that creep about in the sediments by extending pseudopodia from their bodies, feeding on bacteria and other microbes.

Bacteria, fungi, and predatory protists are not usually considered wildlife, since most of them are too small to be seen without using a microscope. Nonetheless, these single-celled organisms are vital to the well-being of other creatures in the marsh and estuary. Bacteria and fungi grow on and enrich cordgrass stems and detrital particles that animals can't digest. In turn, bacteria are food for flagellates, ciliates, and amoebae. Both bacteria and protists are good food for the multitudes of tiny animals that live in marsh sediments and in estuarine water. The little animals that feed on the microbes are in turn eaten by crabs, shrimp, fish, and birds, the wildlife that we know and value the most.

But that isn't the end of it. Bacteria, fungi, and protists have another vital job in the estuary: recycling the mineral elements, nitrogen and phosphorus, that accumulate in cordgrass and other plants. The continuous cropping of bacteria by predatory protists keeps the bacteria in a healthy, fast-growing state so that they are able to more rapidly decompose dead plant detritus. Both bacteria and protists excrete simple mineral-rich compounds, including ammonium and phosphate, just like the plant nutrients in garden fertilizer. These mineral compounds are used by algae and marsh plants for growth. In this way, the cycle of growth, death, decay, and new growth in salt marsh estuaries is maintained by invisible but essential microbes.

Bacteria, fungi, and predatory protists are for the most part heterotrophic (other-feeding) microbes; they live on organic matter produced by plants, as animals do. Exceptions to the heterotrophic mode of nutrition and growth are the photosynthetic cyanobacteria and the chemosynthetic sulfur bacteria, which are autotrophic (self-feeding), using either light energy (cyanobacteria) or energy generated from the reaction of sulfide and oxygen (sulfide-oxidizing bacteria), as already mentioned. Also, in the case of predatory protists, some are able to capture chloroplasts, light-harvesting organelles, from their algal prey and then use them to photosynthesize. These are referred to as "mixotrophic" protists, because they mix together the trophic modes, or ways of gaining food, of animals and plants.

The other group of microbial autotrophs in the sea is composed of single-celled algae. These eukaryotic cells use energy from sunlight (true solar cells!) to create organic matter from carbon dioxide. Algae have light-harvesting chlorophyll pigments in special organelles, chloroplasts. Chloroplasts are the basis of photosynthesis in all land plants as well. The ultimate origins of these organelles were symbioses established between ingested cyanobacteria and predatory protists, perhaps as long as two billion years ago. All chloroplasts still carry a bit of bacterial DNA from the original, free-living cyanobacterium that was captured and kept by a predatory protist. All along the evolutionary history of protists there have been instances when a cyanobacterium, or a chloroplast-bearing alga, has been ingested but not digested by a predatory protist, resulting in the protist acquiring a symbiotic chloroplast that ever after provided it with a photosynthetic lifestyle. (These may be the only instances of a truly free lunch!)

For the most part, present-day algae do not eat other cells, although there are algal species that can ingest prey or have lost their chloroplasts over evolutionary time and now are fully heterotrophic. Microbial modes of nutrition are complicated, a topic of much study and debate by ecologists. For example, scientists have only recently discovered that many species of chloroplast-containing flagellated algae in the open ocean are capable of feeding on bacteria as a supplemental source of nutrients and trace metals.

Microscopic single-celled algae, which occur in great abundance as phytoplankton in water and as benthic microalgae in marsh muds, are a vital source of food for coastal animals. In most coastal regions, larger, multicelled algae occur as filmy or leafy plants attached to rocks, pilings, and any other hard surfaces. The southeastern coast of the United States has only a few of these macrophytic (big plant) algae because conditions there are not favorable for their growth. The turbidity of the estuary doesn't allow sufficient sunlight to filter down to the sediments; the shifting mud and sand make it difficult for macrophytes to attach to the bottom; and the summer water temperatures are too hot for cold-loving big-frond species such as kelp. But some types of macroalgae are able to survive in the salt marsh and in fouling communities (collections of encrusted organisms, such as mussels and barnacles, and their predators, such as crabs and starfish) on pilings; these algae are most obvious during the winter.

Although the name "phytoplankton" stems from Greek words meaning "plant wanderers," these microscopic algae in the plankton are not really plants in the sense that grass and trees are. Analysis of their taxonomic relationships by molecular genetics shows that marine phytoplankton are highly diverse, with many deep evolutionary divisions. Only one group of algae, the green-pigmented chlorophytes, is directly related to land plants. Most other marine algae have additional reddish-brown or golden pigments and don't appear green in color.

Algae suspended in estuarine waters grow quickly, nourished by nutrients brought down from rivers and by microbially generated ammonium and phosphate from decomposing cordgrass detritus. In turn, predatory protists and tiny animals in the water, the zooplankton, capture and eat the phytoplankton and then are themselves eaten by bigger creatures. Benthic-dwelling filter feeders—mussels, oysters, and clams, along with members of the dense fouling communities growing on floating docks, pilings, and other surfaces—strain both phytoplankton and zooplankton from the water.

All over the ocean, wherever there are abundant mineral nutrients for algal growth, one group dominates the phytoplankton: diatoms. Diatoms are essential to marine food webs. There is a quip familiar to biological oceanographers: "All fish is diatom." This means that harvests of cod, salmon, herring, pollock, and every other commercial fish in the sea are ultimately dependent on the mass growth of diatoms.

Diatoms are algae that often grow as chains of cells, easy for small fish and zooplankton to capture. These golden-brown-pigmented algae store excess organic carbon as oils in their cells, so they are a high-energy food for marine animals. Because phytoplankton are often compared to "pastures of the sea," marine animals that eat phytoplankton are called "grazers." The zooplankton that graze on diatoms provide nutritious food for marine fish. Of course, since Pomeroy and other microbial ecologists discovered the roles of bacteria and protists in marine food webs, we know that there is not a simple food chain from diatoms to fish, but rather a complex food web in which protists feeding on bacteria and algae are also eaten by copepods (tiny crustaceans) and other planktonic animals. But recent studies have underscored the fact that diatom blooms are indeed the source of most of the algal biomass that supports the production of fish and shellfish.

Diatom cells are encased in a transparent shell made of silica, which is also the main element in glass. These clear, brittle shells have two halves, called frustules, one slightly larger than the other, so that the two frustules fit tightly together, like the lid and bottom of a box. The frustules are decorated with elaborate patterns of very tiny holes, or pores, through which the diatom takes in nutrients. There are two general forms of these glass shells: centric and pennate. The community of diatoms in the plankton is dominated by centrics. The frustules of centric diatoms are radially symmetrical and usually round or square in shape. Many exist as single cells, such as species of *Coscinodiscus* and *Rhizosolenia*. But other common centric diatoms, for example, the abundant and ubiquitous coastal diatom *Skeletonema costatum*, form long chains of cells linked together by elastic threads. This makes the diatoms even bigger and therefore easier to capture by algal grazers. Some of these chain-forming diatoms, for example, species of *Chaetoceros*, have long silica spines protruding from their frustules, which may deter grazers.

The second major diatom group has frustules that are bilaterally symmetrical; a central shaft has patterns of pores aligned on either side. These are the pennate, or feather-shaped, diatoms. Most pennate diatoms live on surfaces in the marsh mud, on sand grains, on marsh grass stalks, or on any other suitable substrate. These are benthic diatoms. But one pennate species is common in the plankton of the Georgia coast: *Asterionellopsis glacialis*. This striking diatom has a shell shaped like a slender vase. Chains of *Asterionellopsis* cells are often linked together at the vase bottom, forming beautiful circular starbursts—hence the genus name, meaning "starry," of this pennate diatom.

The final group of phytoplankton comprises small phytoflagellates. These unicellular phytoplankton are propelled by one, two, three, or four flagella. They are genetically diverse and contain different types of pigments. Some are green pigmented, like land plants. Many other

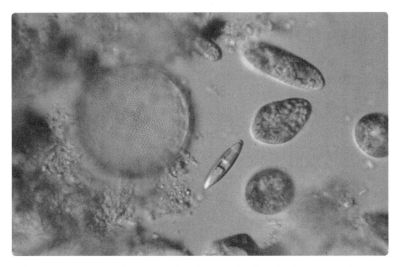

Microscopic algae are important in estuarine food webs: this image shows a large centric diatom (*left*), a smaller, pennate diatom (*center*), and several euglena algal cells (*right*) from moist beach sand.

phytoflagellates have golden-brown pigments similar to those of diatoms. Although these algae are much smaller than diatoms and dinoflagellates, they can grow quickly and make up a large part of the total phytoplankton community.

The mudflats, marsh creek banks, and intertidal beaches of the Georgia coast are often a richly hued golden brown or vivid green. This is visible evidence of the abundance of microscopic benthic algae, which, like phytoplankton in the water above, are a vital food for estuarine animals.

The most common benthic algae are pennate diatoms. Benthic diatoms that live in the marsh mud form mucus-lined tubes, along which they migrate according to the tides. At low tide, the diatoms move up to the marsh surface and form a golden lawn of cells, photosynthesizing in the sunlight. At high tide, when the sunlight is dimmed by estuarine water flowing into the marsh, the diatoms retreat half an inch or more down into the mud. Sheltering in their tubes, the diatoms seek protection from the tidal flow. Even so, clumps of benthic diatoms can be lifted away by strong currents scouring the surface of mudflats. Diatoms grow also on sand grains on the lower beach, and at times are dense enough to be noticed as golden-brown patches on wet sand. Common species of benthic diatoms are in the genera *Navicula*, *Pleurosigma*, and *Nitzschia*. The diatom *Nitzschia paradoxa* occurs as chains of long cells attached side by side by a mucus coating. They move by gliding back and forth along each other. Very large pennate diatoms, such as *Synedra* species, likewise grow as a soft brown "fur" on surfaces and are common in fouling communities.

A bright green color on sediments signals a bloom of green-pigmented flagellated euglenae. Like some dinoflagellates, euglenae have light-detecting eyespots, aids for migrating to optimal environmental conditions. Euglenae can form local patches on estuarine mud. But they reach their maximum abundance when they bloom on the lower intertidal beach in late summer and early fall. During these "green beach" events, when the day is cloudy or in late afternoon, the algae mass on the surface of the sand tints large areas of the beach a vivid color. When the sun is bright and the sand surface hot, the euglenae retreat to moister, shadier subsurface sand layers. But if a towel or an umbrella shades the beach for a few minutes, the euglenae quickly move up to the surface again.

Benthic algae are a basic food for many animals of the Georgia coast. On the mudflats at low tide, crowds of dark half-inch-long mud snails, *Ilyanassa obsoleta*, graze on diatoms. Marsh fiddler crabs, periwinkle snails, and bottom-feeding fish such as mullet consume benthic algae in great quantities. Benthic diatoms and euglenae are a staple of the diet of the burrowing worms, clams, and crustaceans that live in marsh mud and beach sand. In turn, the animals that feed on benthic algae are hunted by blue crabs, shrimp, fish, and shorebirds.

Some simple forms of algae grow as visible flat sheets or filaments consisting of a double layer of cells. Although rare in coastal Georgia, several species of macrophytic (large plant) algae can be found attached to pilings, the stems of marsh grass, and other surfaces. In the winter, a brown filamentous alga *Ectocarpus siliculosus* forms long golden beards on cordgrass stems along the tidal creeks.

A tiny semifilamentous green alga, *Pseudendoclonium submarinum*, is often found on dying cordgrass leaves. It is a food resource for periwinkle snails, which spend most of the summer rasping the surfaces of *Spartina* stems and leaves. Other macrophytic algae are commonly found in the subtidal estuary: green, leafy sea lettuce, species of *Ulva*, and the red filamentous algae *Caloglossa leprieurii*, *Bostrychia radicans*, and species of *Polysiphonia*. *Rhizoclonium*, a small filamentous alga, occurs as a fuzzy green coating on pilings. Sargassum weed, *Sargassum filipendula*, grows in offshore hard-bottom habitats and occasionally washes up on the beach.

Opposite: Intense green bloom of euglenae algae on Nannygoat Beach, September 1983.

3. Marsh Grass, Live Oaks, Sea Oats

Georgia's Golden Isles are named for the expansive meadows of salt marsh bordering the sea islands. In fall, *Spartina* leaves turn a glorious yellow brown in response to cool weather fronts signaling the approach of winter. During spring and summer, the dominant color of the coast is green: the bright green of cordgrass shoots sprouting up from the marsh muds and growing tall along the edges of tidal rivers and creeks, and the dark green of the bushes edging the marsh and of the sheltering live oaks and tall pines crowding the island beyond. The beach dunes sprout green grasses and herbs that will flower pink and yellow in mid to late summer.

In higher plants, the roots, stems, and leaves have internal piping that carries nutrients from roots to shoots, analogous to our own circulatory system. This network of tubes is described as vascular (from *vas*, the Latin word for "vessel"), and higher plants as a group are known as vascular plants. Instead of a central pump, a heart, vascular plants rely on differences in water pressure caused by the evaporation of water from the surface of leaves, which pulls upward a continuous column of water extending down to the roots.

Vascular plants came on the scene long after algae had appeared in the sea. Life on land was slow to get going because plants and animals living in marine habitats had to evolve mechanisms that would allow them to survive without abundant water and while exposed to wind and the sun. By about 450 million years ago, primitive plants related to green algae had colonized rocks lining ocean shores. Lichens, which are symbiotic associations of fungi and either filamentous cyanobacteria or algae, adapted to growing on hot, dry rocks. The fungal hyphae in lichens create the structure of the plant and absorb and store moisture during rainstorms. The algae hosted in the fungal mass produce food by photosynthesis for both themselves and the fungi. Lichens helped break down rocks into soil in which early land plants could grow. These primitive forms slowly spread, developing over time into larger, spore-forming plants, including mosses and ferns.

In the Devonian period, 416 million to 360 million years ago, the land became covered with dense forests of strange tall plants. Primitive *Archaeopteris* trees (long extinct) grew as high as six-story buildings. The first cone-bearing plants with true seeds, the gymnosperms, appeared during the Devonian. During the Jurassic, when early dinosaurs roamed, early gymnosperms such as dawn redwoods and cycads proliferated; both are still around today. On Sapelo, a big cycad, with a palmlike trunk and long fronds surrounding a large yellow pineapple-like seed head sprouting from the crown, grew beside the tennis court near the Marine Institute. The flowering plants, angiosperms, evolved during the late Jurassic and Cretaceous periods, when dinosaurs ruled the continents. These plants, which have seeds enclosed in a hard cover or fruit, were better adapted to cold and drought.

Modern vascular plants are divided into two major groups: the cone-bearing gymnosperms, such as pine trees, which have naked seeds, and the flowering angiosperms, which have seeds protected in a capsule. In turn, there are two major groups of flowering plants: the monocotyledons (monocots), whose germinating seeds have only one (mono-) cotyledon, or initial leaf (corn is an example); and the dicotyledons (dicots), which have two initial leaves when germinating (beans, for example). Another obvious difference between these types of flowering plants is that monocots have leaves with parallel veins, while the leaves of dicots have a network of veins (compare a grass leaf with the leaf of an oak).

A special group of flowering plants, grasses, did not exist during the time of the dinosaurs. Grass pollen first appears in the fossil record only about 55 million to 60 million years ago, long after the major extinction event that ended the era of reptiles and gave the green light to the evolution of mammals. Sweeping grasslands that encouraged

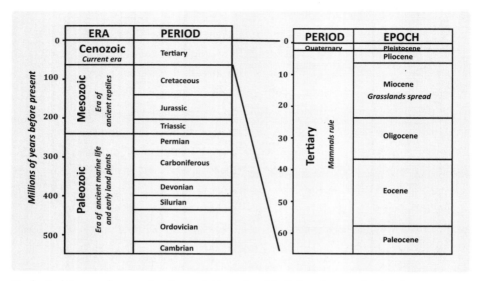

Geological time scale spanning the evolution of multicellular plants and animals.

the evolution of swiftly running grazers such as modern horses and antelopes became extensive in the Miocene epoch, 23 million to 5.3 million years ago. Humans have a special relation to this widespread group of monocot plants. Most of our major food crops, including wheat, corn, oats, rice, and sugarcane, are grasses. Wheat was the earliest domesticated plant. Inhabitants of small villages in Asia Minor (modern Turkey) were sowing and harvesting emmer wheat, a big-seeded grass that grows wild in the Middle East, over 9,000 years ago.

Grasses are hardy and can thrive in both cold and hot climates. Some species of grass have adapted to salty habitats too. Grasses that grow in the salt marshes and in the beach dunes have a special mechanism for fixing carbon dioxide. This was first discovered in sugarcane, a large fast-growing tropical grass. The c-4 pathway of photosynthesis enables plants to convert carbon dioxide into sugars very efficiently, with little water loss from the leaves, and it operates well at high temperatures. This gives marsh grasses a competitive edge over species that only have the regular, or c-3, pathway for fixing carbon (there is more about this in chapter 15).

Several types of grass are abundant in the marshes and beach dunes of coastal Georgia. Species of dicot plants, which include herbs, bushes, and trees, are much more diverse but not nearly as striking as the vast expanses of cordgrass immortalized by Sidney Lanier in his poem "The Marshes of Glynn." Sea grasses, the only seed-bearing land plants that have recolonized the sea, grow completely underwater, rooted in shallow sandy sediments. They do not occur along the Georgia coast, since the estuarine water is too murky to allow enough sunlight to filter down to the bottom at high tide, and the mudflats and sandflats are exposed at low tide. Sea grass beds are common to the south along the Florida coast, and farther north, from North Carolina to Maine, where they flourish in clearer waters with less tidal range. (Sea grasses, although they are monocots and certainly look the part, are in fact not true grasses.)

Plants living in the marshes and on the dunes have adapted to harsh conditions of salty water and soil, high summer temperatures, and water loss through the leaves, which can lead to desiccation. Each plant species has evolved ways of handling these problems. Some, like *Spartina*, have special salt-secreting glands on their leaves, and others have succulent leaves that resist drying out or that store excess salt in the leaf tissues; the salt-laden leaves eventually drop off the plant. Some, again like *Spartina*, have the efficient C-4 biochemical pathway, which allows the plant to take in carbon dioxide from the air with minimum evaporative loss of water.

Now that we know something about the vascular plants, we can survey native vegetation along the Georgia coast. Plants may seem boring; they just sit there rooted on the spot, not doing much other than desultorily rustling their leaves in the breeze. But for a nature lover, plants have a decided advantage over microbes: you can easily see them. And unlike wild animals that flee when you approach, you can reach out and touch plants. Feel the leaves—smooth or hairy, waxy or tender? Crush the leaves in your fingers and inhale a grassy or pungent odor. Closely examine the flowers, perhaps with the aid of a hand lens. The structure of a flower can be used to identify the family of its parent plant.

Salt Marsh Plants

Let's start with the salt marshes. *Spartina alterniflora* is the grass that covers these verdant muddy plains. Although cordgrass is only one species, its plants have variable success in growing, from lush stands six to eight feet high along the creek banks to swards of stubby plants less than a foot tall in the upper reaches of the marsh. Like lawn turf, cordgrass has an extensive system of underground roots and rhizomes, from which new shoots emerge as the rhizomes extend through the sediments. It is difficult to dig into the salt marsh because the tough rhizomes are so dense. The thick, tangled underground growth supports the muddy soil, so it is easy to walk over the marsh plain without sinking in. But close to the creek banks, watch out. The rhizomes of tall *Spartina* plants are less dense, and the soft mud of the creekside marsh is a real foot sucker. I have lost more than one sneaker in mucky sediments near a tidal creek.

When a chunk of soil full of rhizomes is dug up, the soil surrounding the live roots is a bright orange red. This color results from the oxygen that filters out of the roots and oxidizes the abundant iron in the anoxic muds, turning it into the same compound found in rusted metal. The clay soils of the Georgia foothills, which are brought to the coast by river flow and form estuarine mud, are likewise red because of oxidized iron. When the mud becomes saturated with seawater and cut off from the air, the iron loses its oxygen—in effect, unrusting and turning as black as an iron skillet.

We know that *Spartina* copes with the stress of salt in tidal water and marsh muds by excreting it from glands in the leaves. On a hot summer day, salt crystals sparkle on the sides of the shoots. Cordgrass has another problem: it grows in anoxic marsh sediments, but the roots need oxygen to live. This difficulty is solved by the hollow stems of *Spartina*, which funnel oxygen from the air down to the rhizomes and roots in the anoxic mud. An additional problem caused by marsh mud is not so easily overcome. Marsh soils are full of toxic sulfide, produced by sulfate-reducing bacteria living on the abundant sulfate in seawater brought in by the tides. Sulfide can kill delicate plant roots.

Chuck Hopkinson, then on the research staff of the Marine Institute, wading in a mucky creekside cordgrass marsh around Sapelo Island, summer 1985. A large raft of dead *Spartina* stems carried in by high tide blankets the cordgrass at the upper right.

Experiments have shown that the growth of *Spartina* is inhibited by high concentrations of sulfide around its roots. In fact, the black marsh sediment reeks of a rotten-egg smell resulting from sulfide. Along with salt stress, large amounts of sulfide in higher-elevation soils contribute to the stunting of marsh plain plants.

Cordgrass stems, which are more fibrous than the leaves, are the last parts of the plant to decay. In fall, the long stems of dead creekside *Spartina* are washed out of the tidal creeks by extreme high tides, forming great rafts of gray stems that drift around the estuary. These rafts are often carried up onto the marsh, killing the *Spartina* plants beneath them. A high tide that later lifts the stem rafts away will reveal a patch of mud in the marsh, which will stay bare until the rhizomes below sprout new *Spartina* shoots. Dead stems can also drift out of the estuary to the open ocean on the tides, and then wind up far offshore or washed up onto the beach. Although the rafts are impressive, they are only a small part of the total dead mass of *Spartina*, most of which decays into small particles.

Another species of *Spartina* lives in less salty habitats along coastal rivers. In the lower Altamaha and Savannah Rivers, where estuarine water is fresher than in the barrier island marshes, grow stands of giant cordgrass, *Spartina cynosuroides*. These resemble the most robust smooth cordgrass plants. Small marshy islands in the river mouths are often covered with giant cordgrass. In early spring, when most of the cordgrass leaves are still brown and dry, it used to be common practice to burn these islands in order to promote the growth of new cordgrass as well as to kill the larvae of biting flies, which live in the marsh mud.

Besides the swaths of *Spartina*, the other obvious feature of southeastern salt marshes is the silver-tipped black needlerush, *Juncus roemerianus*. This nongrass monocot grows in dense, dark gray-green patches from three to six feet high in the upper marsh. Needlerush is a well-deserved name: the stiff, tubular leaves, hard and sharply pointed at the ends, can deliver a painful prick. *Juncus* excretes excess salt into the tops of the leaves, which then die and lose their color, giving a silvery cast to stands of this rush.

Away from the expanses of stubby cordgrass and clumps of tall black rushes is the low, scraggly vegetation at the landward edge, where the marsh is flooded only during the highest tides. The sandy soils are too salty for *Spartina*, but other plants can survive. Here are two grasses even hardier than cordgrass: salt grass, *Distichlis spicata*, and dropseed grass, *Sporobolus virginicus*. The two species are difficult to tell apart, and both are similar to the Bermuda grass common in southern lawns. Salt and dropseed grasses have narrow leaves that retard water loss, which helps them withstand the harsh conditions of the marsh edge. And like *Spartina*, they have the very efficient C-4 pathway of carbon uptake.

Salt barrens at the edge of the marsh, where salt grasses and succulent plants grow. The small dots in the wet part of the barrens are sand fiddler crabs feeding during low tide.

Patches of succulents with thick fleshy stems are mixed in with the high marsh grasses. These are dicot flowering plants that cope with this salty, dry habitat by secreting excess salt taken up by their roots into the tips of their stems or leaves, which then turn red, die, and drop off with the unwanted load of sodium chloride. Glassworts, *Salicornia virginica* and *Salicornia bigelovii*, are low-growing fleshy plants. The leaves of glassworts are just tiny scales on the thick, segmented stems. Because of the salt accumulated in glasswort stems, coastal residents have long used *Salicornia*, fresh or pickled, as a piquant addition to their meals. (I have found bright green *Salicornia* stems, labeled "sea beans," for sale in the produce section of a grocery store in Oregon.) When the weather turns cool in the fall, the whole plant dies back, and patches of glasswort turn the upper marsh a cheery red. Saltwort, *Batis maritima*, is another common succulent of the upper marsh. Saltwort can be distinguished from *Salicornia* by the boxy, bright green leaves, which grow in clumps on woody brown stems. Saltwort plants and their seeds are nutritious and are cropped by island deer.

Moving from the upper marsh to dry land: bordering the tidal marsh is a hardy community of dicot plants that are tolerant of salt and thrive in full sun. Members of the composite, or daisy, family are adapted for growth in harsh, bright conditions; it is no surprise that many of these plants live at the marsh edge. The flowers of this large family have a surprise: each blossom is a composite (hence the group name) of many small separate flowers of two types. Close inspection of a daisy or sunflower reveals flowers shaped like petals, the ray flowers, arrayed around the edge. In the crown of the composite bloom, a multitude of small, tubular disc flowers crowd together. The composite's many-flowered nature is revealed when the seeds mature. In sunflowers, the hard seeds are packed together where the disc flowers bloomed. In mature dandelion flower balls, each seed is topped with a white tuft; with a quick breath, you can send the seeds floating away under their own individual parachutes.

The sea oxeye, *Borrichia frutescens*, an aster with thick leaves and woody stems, grows as a small bush a foot or so tall. The yellow, daisy-like flowers that bloom in summer form brown burrs on the ends of the stems by fall. A coastal resident told me that her mother used to give her tea made from dried sea oxeye leaves when she suffered from a cold. The drink was so bitter that she willed herself to be well in order to avoid having more of the stuff forced on her. Interspersed with sea oxeye are two other bushy composites. Marsh elder, *Iva frutescens*, has narrow leaves and spikes of small white flowers. It tolerates salt, but not having its roots completely submerged in water. This aster lives only above the high tide line, thus earning the common name high tide bush. Groundsel, *Baccharis halimifolia*, is similar in form to marsh elder, but has wider, more deeply indented leaves. In the fall, groundsel bushes, like dandelions, produce abundant cottony-white tufts that disperse their seeds on the wind.

Down among the composite bushes at the marsh edge, in a whorl of dark green leaves, is a small herb, sea lavender, *Limonium carolinianum*. Also called marsh rosemary, sea lavender is not related to either true lavender or rosemary, but instead is in the leadwort family. In summer, branching stalks bearing delicate purple flowers shoot up from the basal leaves. The ability of sea lavender to withstand salt spray makes it popular with coastal gardeners.

Island Plants

In traveling from the marsh to the dune side of an island, one passes shrubs and trees that have colonized the island interior and the hammocks lining the tidal rivers. These deserve a mention, since they shelter birds and animals that frequent the salt marshes at low tide.

The most common shrub of the Georgia coast is wax myrtle, *Myrica cerifera*. Wax myrtle grows in thick stands along the borders of the salt marsh and on approaches to the beach dunes. When crushed, the shiny dark leaves have a pleasant smell. In fall, tiny bluish-gray berries crowd the branches; their waxy coating is used to make bayberry candles. Flocks of migrating myrtle warblers stop at the end of the summer to feast on the lipid-rich fruits.

Often mixed in with myrtles is a low-growing tree, juniper or red cedar, *Juniperus virginiana*. Although not a true cedar, junipers are related gymnosperms, easily recognized by the bushy twigs covered with tiny, scalelike leaves; abundant blue cones resembling berries; and shaggy gray bark covering the aromatic red wood.

The other gymnosperm trees growing around the marshes, and at times well out into the dunes, are southern yellow pines, including loblolly pines, *Pinus taeda*; slash pines, *Pinus elliottii*; and longleaf pines, *Pinus palustris*. These hardy pines dominate southern tree farms. Seedlings have dense crowns of needles that render them especially adapted to survive the fires that often sweep through coastal forests. The original forests of the Georgia sea islands were stands of large pines with an understory of live oak. According to Wilbur Duncan in his monograph *The Vascular Vegetation of Sapelo Island*, the removal of the tall old pines by early settlers allowed the live oaks to grow and dominate the forest canopy.

Two plants used in folk medicine can be found along marsh edges and in back dunes. Yaupon, *Ilex vomitoria*, is an understory shrub related to the holly, bearing small glossy leaves and single red berries. Infusions of the leaves can be used as an emetic, hence the species name. One of the more striking coastal plants is Hercules-club, *Zanthoxylum clava-herculis*, with a thick stem from one to two feet tall, up to the size of a small tree. This devilish-looking plant has large leaves divided into four pairs of leaflets, with spiky thorns on both leaf and stem. As if the thorns weren't enough protection from browsers, the leaves produce a compound that anesthetizes an incautious grazer's mouth. Native Americans and settlers used these leaves to relieve toothache pain.

Native palms are abundant on Georgia sea islands. Cabbage palm, *Sabal palmetto*, a tall tree with large fan-shaped leaves, often borders salt marshes. At the center of the leaf cluster is the tender heart of palm of gourmet salads, although cutting out this growing bud kills the tree. Saw palmetto, *Serenoa repens*, grows in low dense thickets up to the edge of the marsh. Saw palmetto has large fan-shaped leaves that sprout from thick, branched stems lying horizontally on the ground.

The best-loved coastal tree is the live oak, *Quercus virginiana*. The low spreading branches, draped with Spanish moss, form an iconic hallmark of the South. Live oaks are home to innumerable other plants, termed epiphytes, since they live on the surface (epi-) of the oak branches. Spanish moss, *Tillandsia usneoides,* is not a true moss, but instead a bromeliad, a flowering plant related to pineapples. The tangled mass of stems bearing scaled leaves hangs down from the trees like a gray beard and provides shelter to many insects, spiders, snakes, and even bats. A species of jumping spider, *Pelegrina tillandsiae,* which has a body camouflaged to mimic the color of the *Tillandsia* stems, is endemic in Spanish moss. Pink lichens grow on the thick, corrugated bark of the oak trunks. The huge drooping limbs of the trees host a garden of resurrection ferns and vines. Once I was shown a rather large prickly pear cactus growing in the crotch of an old oak. Warblers are fond of raising their young in the big oak trees, using Spanish moss to line their nests. On Sapelo Island, each oak seemed to have its resident family of gray squirrels. Island deer also fatten up for the winter on the abundant crop of acorns produced by live oaks each fall.

Unlike oaks that lose their leaves in autumn, live oaks keep green leaves year-round. But that doesn't mean the leaves don't drop, as I found out each April when we had to sweep the porch of our house every day to keep up with falling oak leaves. Each leaf stays on the tree for only about two years. During spring, live oaks shed half their leaves, the ones that are two years old, and grow tender new leaves to replace them.

The other bit of live oak lore I learned was that during the age of sailing ships, live oaks on Georgia sea islands supported a thriving business: the export of naturally curved, strong hardwood to shipbuilders in the Northeast. Timber from live oaks on St. Simons Island was used for internal framing in the first American naval frigates. The most famous of these is the USS *Constitution,* constructed in 1797. The Georgia live oak wood in the hull was so tough, and so impervious to British cannon fire in the War of 1812, that the ship was given the famous nickname "Old Ironsides." The USS *Constitution* survived the war, and has been preserved as a historical museum at the Charlestown Navy Yard in Charlestown, Massachusetts.

Dune Plants

At the seaward end of an island road lie the beach and its dunes. After the point where the myrtle bushes thin out, the dunes farthest from the ocean harbor small meadows of grasses and flowering herbs. Many of the salt-tolerant plants bordering the salt marsh also grow around brackish-water sloughs that collect in the hollows between the tall back dunes. Hardy pines or live oaks growing in the back dunes often have a windswept look, since the more luxuriant foliage on the branches faces away from the ocean. This iconic form of coastal tree owes its appearance to the pruning effect of salt on the twigs most exposed to wind-whipped sea spray during storms.

Be careful walking here! Large diamondback rattlesnakes prowl the dune meadows. There are also two types of prickly plants that hide in the sand and give unexpectedly painful jabs to dune hikers.

The dune habitats offer one more species of cordgrass: salt hay, *Spartina patens*, a bushy grass with long narrow leaves, unlike the other species of *Spartina*. In New England salt marshes, salt hay grows in extensive stands and was once used as fodder for cattle. In southeastern marshes, salt hay is found only in isolated clumps along the edge of the cordgrass marshes or at the edges of sloughs in the beach dunes.

In late summer, large tufts of muhly grass, or sweetgrass, *Muhlenbergia capillaris*, produce feathery pink seed heads that lend a delicate beauty to the back dunes. This grass is used in traditional basket making along the coasts of Georgia and South Carolina. Employing a technique brought from Africa, Gullah and Geechee artisans form coils of dried sweetgrass stems and bind the coils together with strips of palmetto frond to build a sturdy frame for their baskets. When I was living on Sapelo, there was a master basket maker, Allen Green, in the island's African American community. One of Mr. Green's specialties was a "fanner," a wide, flat basket with a rim. In the past, when rice was grown on the sea islands, fanner baskets were used to "fan," or toss, rice grains in the air to separate the dry chaff from the kernels.

Fanner basket made by the master artisan Allen Green of Sapelo Island. It was constructed from coils of dried sweetgrass stems bound together with strips of palmetto leaves. In the past, this type of basket was used to separate, or "fan," the chaff from rice grains.

A most striking plant of the dune habitat is Spanish bayonet, *Yucca aloifolia*. A member of the lily family, the yucca has large, succulent, sharply pointed leaves and grows several feet tall. Tall spikes of white bell-shaped flowers bloom in summer. These big lilies were introduced to Sapelo by Thomas Spalding during the plantation years to form hedges around fields of sea island cotton. Since then, *Yucca* has become naturalized all over the island, including in the sand dunes. I was always impressed with the size of this lily's leaves and flowers. (Until, that is, I traveled to Australia and encountered the gymea lily, *Doryanthes excelsa*, the largest lily in the world, a veritable Godzilla of lilies, with leaves as tall as a person and a flower stalk twenty feet high, topped with a red flower cluster a foot in diameter. For me, at least, that put the yucca's flower spike in perspective.)

Also deserving special mention is the prickly pear, *Opuntia*, the only cactus native to the Southeast. The familiar species *Opuntia humifusa* is easily recognized by its flat fleshy leaves, small thorns, and beautiful waxy yellow flowers. It grows in largish clumps around salt marshes and behind the beach dunes. In summer, the purple pear-shaped fruits can be collected and eaten or made into wine-colored jelly (after the thorns have been carefully removed!). The other species of prickly pear is *Opuntia pusilla*, creeping cactus, a nasty little succulent that hides in the grass in the back dunes. The small round segments are covered with very long, sharp thorns; these segments readily detach and stick into anything that brushes by.

Closer to the beach is a composite bush, *Iva imbricata*, a close relative of the marsh elder. Unlike its marsh cousin, the seashore elder has thick leaves and chunky stems. Other salt-tolerant composites grow here too. Seashore goldenrod, *Solidago sempervirens*, brightens both the marsh edge and dune meadows with golden flowers in late summer. Camphor weed, *Heterotheca subaxillaris*, a low-growing composite, has small tough leaves. Crushed, the leaves emit the distinctive pungent odor that gives this plant its common name. This inconspicuous little composite, which grows all over the dunes, comes into its glory in summer, when it is covered with golden-yellow flowers.

Small herbaceous plants are abundant on the dunes. Many have close relatives adapted to living in deserts. These beach herbs have special ways of dealing with the problems of salt and desiccation. Sea rocket, *Cakile edentula*, is a thick-stemmed plant with indented fleshy leaves. A member of the mustard family, sea rocket is edible and adds a spicy tang to salads. Its white flowers have the shape characteristic of mustard plants: four equally sized petals and four long stamens arranged in a cross. The mustard family, Brassicaceae, is also called Cruciferae (Latin for "cross bearing"), which refers to the cross-shaped flower parts. It is not surprising that sea rocket is good to eat, since many common vegetables, including cabbage, broccoli, kale, and radishes, are also in the mustard family.

Two other dune species are euphorbs, or spurges, a family of succulent, drought-resistant herbs. Beach croton, *Croton punctatus*, is a grayish dune plant that grows as a low round bush; it has fuzzy oval leaves and small white flowers that produce three-sided fruits in the fall. Beach spurge, *Euphorbia polygonifolia*, a cousin of beach croton, can be distinguished from its relative by its small tapering leaves and milky sap. Saltbush, or orach, *Atriplex cristata*, is a low-growing herb with small flower spikes and small triangular leaves covering the stems.

In the first line of dunes, you might notice a low spreading bush with many stems bearing tiny leaves and sprouting from a central root. This is Russian thistle, *Salsola kali*. Tiny pinkish-purple flowers appear on the stems during summer. In the fall, the plant dries out and may be blown loose down the beach. This is, in fact, the tumbleweed of cowboy movies. Russian thistle, native to the arid steppes of the Ural Mountains, was accidentally introduced into the United States in the 1870s. This invasive weed has since spread throughout dry plains and deserts to become an iconic (and the same time ironic, since it is

The author's son Aaron posing with sea oats in the fore dunes of Nannygoat Beach on Sapelo Island, 1984. Sea oats and other grasses stabilize the beach dunes.

Marsh Grass, Live Oaks, Sea Oats

a Russian immigrant!) symbol of the American West. The genus name *Salsola* derives from the Latin word for salt, indicating the ability of this plant to tolerate salty as well as dry habitats.

Some plants are notable for their low, creeping growth over the dunes. Beach pennywort, *Hydrocotyle bonariensis*, has circular, bright green leaves that push up through the sand from thick white rhizomes below. Pennyworts grow at the edges of beaches and marshes all over the world. Another creeping plant is beach morning glory, *Ipomoea imperati*, a white-flowering vine related to the sweet potato.

The dunes that front the beach are colonized by several kinds of grasses. The most impressive dune plant, sea oats, *Uniola paniculata*, stabilizes the dunes with its extensive root system. Sea oats are so vital to keeping the dunes intact that in Georgia they have legal protection.

Among the tall sea oats is another grass with broad green leaves. This is panic grass, *Panicum amarum*. Its leaves are similar in shape to those of sea oats, except that they are not as curly and grow out alternately from the stem. Mid to late summer sees long seed spikes on *Panicum* and big-seeded stalks on sea oats. The presence of a usually inconspicuous dune grass, the well-named sandspur, *Cenchrus tribuloides*, is also announced by its seeds. These are most painfully obvious when the sharply spiked seeds, larger that those of the common sandspur, jab into a bare foot. Our family called these prickly devils "land sharks."

Plants, from microscopic algae in the water and on the mudflats to the trees of the marshes and sea islands, underlie the food webs that support the animals living along the Georgia coast. While some coastal animals are well known, most of them are easily overlooked. But all play vital roles in coastal ecosystems, and a few are really odd.

Creatures of the Black Goo 4

Our border collie, Yofi (a Hebrew word meaning "isn't that nice?"), had a pal, March, a medium-size brown mutt who was a neighbor's dog. In the morning, a yapping at our porch door meant that March was outside, asking Yofi to come out for an adventure. The two would disappear down a dike running along the marsh fronting our backyard and then would often be gone all day. When they reappeared at dinnertime, they were usually indistinguishable, both a dark chocolate from the marsh muck covering them from nose to tail. When we washed our pup off, we knew that in all that mud streaming off her fur were not just microbes but all sorts of small animals that inhabited the black goo of the marsh soils and mudflats.

Although Georgia marshes and estuaries don't have the great biodiversity found in rocky intertidal communities or coral reefs, the animals that thrive on Georgia coasts occur in teeming abundance. Most are invertebrates, creatures without a spine or an internal skeleton. Of the recognized major categories, or phyla, of animals, virtually all comprise creatures with invertebrate body plans. All vertebrates, including fish, amphibians, reptiles, birds, and mammals, are included in a single subgroup of the remaining phylum, Chordata. One type of invertebrate is grouped with the vertebrates. Squishy tunicates, or sea squirts, have a fish-like larval stage with a primitive backbone, and so are included with the chordates. Hemichordates, or acorn worms, which live in burrows in sandy marine sediments, also share some chordate characteristics.

There is a curious embryological difference between the chordates and members of the other animal phyla. When a fertilized egg begins to divide, a small sphere of cells, the blastula, initially forms, with

Extensive mudflat at low tide in Barn Creek on Sapelo Island, summer 1974. Marsh muds harbor myriad tiny animals that are food for fiddler crabs, snails, and birds.

an opening, or pore, at one end. For all invertebrates other than sea squirts and acorn worms, the initial embryonic pore develops into the mouth. For the chordates, including us, the pore develops into the anus. (So we humans all start out as . . . hmmm.)

"Meiofauna" (Greek for "small animals") is the scientific term for the multitude of tiny creatures, smaller than a poppy seed, that live in marine sediments. Meiofauna include single-celled predatory protists, tiny worms, and minuscule crustaceans. These feed on bacteria and algae in the mud, and are themselves food for larger benthic-feeding animals. Sediments contain many different microhabitats: for example, coarse-grained sandy sediments are porous, allowing nutrients and oxygen from the overlying water to filter deep into the subsurface, while silty sediments made of compact, fine particles don't permit the diffusion of oxygen below the upper few tenths of an inch. Intertidal marsh soils in which plants grow offer habitats different from tidal creek bottom muds full of decomposing cordgrass detritus. Because of the great variety of habitats, the species diversity of the meiofauna is much greater than the diversity of the tiny creatures, the zooplankton, in the estuarine waters that periodically inundate the marsh.

The most numerous groups, in abundance and number of species, in the meiofauna are unicellular protists: microscopic flagellates and ciliates. Unlike the single-celled plankton in the overlying water, these protists are well adapted to life among sand grains and clay particles. Most are long and thin, which allows them to squeeze through tight spots between mineral grains. Some of the ciliates have stiff hairs, or cilia, on their undersides to aid in scrambling over minuscule sand boulders or in scraping algae off the grains. When viewed live under a microscope, these benthic ciliates scurrying about on their stiff cilia have an eerie resemblance to cockroaches.

Worms are the most abundant animals in the meiofauna, and nematodes, microscopic roundworms, are by far the most numerous. In fact, nematodes are the most abundant type of multicellular animal on the planet. It has been said that if all other living tissue, including all of it in plants and soils, were to vanish, the outline of the biosphere would still be defined by a squirming mass of these tiny worms. Most nematodes are less than one-eighth of an inch long. They move through sand and mud with constant whiplike motions of their bodies, like microscopic snakes. Different nematode species specialize in feeding on organic detritus, bacteria, algae, living plant tissues, or other small animals in the sediment. Not very much is known about the importance of nematodes in marine habitats, although in marsh mud the total mass of nematodes often rivals the mass of bacteria, which is significant. Nematodes are probably a staple in the diet of many benthic-feeding animals.

Polychaete worms, marine worms with many stiff hairs on their segments, generally are too big to be classed with the meiofauna. An exception is a tiny salt marsh polychaete, *Manayunkia aestuarina*, which is so abundant that it makes up a sizable part of the meiofaunal animals until it grows large enough (though still no longer than an quarter of an inch) to be classed with more substantial fauna. This light brown segmented worm has eight yellow tentacles around its mouth, which are used to grab bits of food and to build the miniature mud tube in which the worm hides.

Tiny crustaceans form a second abundant type of meiofauna in the sediments. Crustaceans are a major division of the phylum Arthropoda, which includes all joint-legged animals with a hard exoskeleton made of chitin, a nitrogen-rich carbohydrate polymer that forms a rigid covering for the soft tissues within. The downside to having an external cover is that marine crustaceans must shed, or molt, their old shell and make a new one as they grow larger.

Crustaceans evolved in Cambrian oceans hundreds of millions of years ago. Trilobites, now extinct, and horseshoe crabs, still with us and essentially unchanged from those primordial times, were among the first invertebrates to have the advantage of a rigid coat of armor. Later, marine arthropods adapted to life on land, and their descendants, insects and spiders, have been wildly successful, colonizing every imaginable terrestrial habitat.

Although distant relatives of the tiny crustaceans in the zooplankton swimming in the water above, meiofaunal crustaceans, which are adapted to living on surfaces, are placed in separate taxonomic groups. The most common of these crustaceans are the harpacticoid (from the Greek for "rapacious") copepods, active little animals with mouthparts that rake tasty tidbits off clay particles and sand grains. Because there are so many different habitats in the estuary and in near-shore sediments, there are many more species of benthic dwelling harpacticoid copepods than there are calanoid copepods in the zooplankton. For example, in a ten-year study in a South Carolina salt marsh estuary, Bruce Coull and Bettye Dudley, scientists at the Belle W. Baruch Institute for Marine Biology and Coastal Sciences, documented seventy-three species of harpacticoid copepods living in a muddy habitat and fifty-six species in a sandy site. Some of the species were adapted to living in the sediment, and some at the sediment surface. Some species had food preferences for bacteria-rich organic detritus or for benthic algae and protists.

Another type of miniature crustacean, the ostracod *Cyprideis floridana*, is a common inhabitant of salt marsh soils. Ostracods are funny little creatures that are also called mussel shrimps because their bodies are encased in two oval chitin shells that resemble the shells of bivalves.

Animals living in and on the sediments and larger in size than a poppy seed, and thus visible to the naked eye, are called macrofauna (Greek for "large animals"). As with the meiofauna, the macrofauna are dominated by worms, worms, and more worms, with arthropods in second place.

Annelid worms are incredibly numerous in marine sediments. Species of oligochaetes (Greek for "few long hairs"), earthworm-like annelids, are found in salt marsh soils and in beach sands. Some of these small marsh worms have discovered a comfortable haven above the mud. These specialized oligochaetes live in spaces in the sheaths of *Spartina* leaves, where the leaves grow out from the stems. Here the worms feed on live cordgrass tissue and bits of detritus brought in by the tide, and are protected from predators in their leafy refuge.

The vast majority of marine worms, and the most spectacular, are the polychaetes ("many long hairs"), or bristle worms. The diversity of marine polychaete worms is staggering. Bristle worms are named for the many fine hairs, or setae, projecting from each body segment, which aid the worm in moving about, in, or over the sediments. Many polychaete worms are bright red in color from the iron-rich heme compounds in their blood, which bring oxygen to the worm's tissues just as hemoglobin does in humans.

The common clam worm, *Alitta succinea*, a bristle worm several inches long with a body colored variously from yellow to red, is ubiquitous in southeastern estuaries. The clam worm is not picky about type of bottom (mud or sand), salinity, or diet, and is found all over the estuary—in fact, in intertidal habitats around the world. *Laeonereis culveri*, a cousin of the clam worm, prefers muddy sediments and is abundant in marsh soils ranging from soft creek-bank muds to the salty flats in the upper marsh. Other polychaetes are found in both mud and sand. These include the red-lined worms in the genus *Nephtys*, which have gray bodies with distinct blood vessels running down the back and abdomen; the opal worm, *Arabella iricolor*, with a long, slim, brilliantly iridescent body; the carnivorous bloodworms of the genus *Glycera*, which have tiny black fangs they use to capture prey; and the threadworm *Drilonereis longa*, which looks just like a long red elastic thread.

These polychaetes are errant worms; that is, they roam freely through sand and mud in search of microbe-rich detritus, algae, or other morsels to eat. Even more interesting are the sedentary bristle worms, which establish burrows in the sediments, emerging to capture tiny creatures in the water passing by or on the nearby surface and then retreating to a safe depth to avoid bottom-feeding fish and crabs. Several common burrowing polychaetes in Georgia estuaries line their holes with an organic tube. *Kinbergonuphis microcephala* builds a smooth, parchment-like tube that projects up from the surface of sand. The tops of these tubes can often be found washed up in great numbers on the beach. Because getting enough oxygen is a challenge for a benthic-dwelling animal living in an underground burrow, the body of *Kinbergonuphis* is crowned with red filamentous gills, which scavenge oxygen from seawater that the worm pumps through its tube.

While *Kinbergonuphis* worms are usually found in shallow parts of the estuary and near shore, another burrowing polychaete, the bamboo worm *Clymenella torquata*, makes a similar tube in deeper-water sediments. This bristle worm gets its common name from the resemblance of its segmented body to the jointed stems of bamboo. The plumed worm *Diopatra cuprea* is a close relative of *Kinbergonuphis*. The tube of *Diopatra* is unmistakable because the worm glues bits of shell and other debris to the top of the tube, which projects from the sand. The plumed worm gets its name from the bushy red gills that decorate the top third of its foot-long, iridescent blue body.

The burrows of one of the most spectacular coastal bristle worms are made of golden thread. The magnificent scale worm *Polyodontes lupinus* builds silky tubes in mud along protected beaches. The body of the scale worm, covered with small scales, can reach three feet in length. The worm spins gold-colored filaments from each body segment, using them to construct its soft home in the muddy sand. More modest dwellings are built by the inch-long sand tube worm *Sabellaria vulgaris*. Its tubular houses are made of sand grains cemented together; the houses are fastened to shells or to one another in little worm "cities."

Piles of stringy mud castings on sandy mudflats mark the burrows of the acorn worm *Saccoglossus kowalevskii*. This large worm is not a polychaete but a hemichordate, sharing some features of sea squirts and vertebrates. Digging an acorn worm out of its U-shaped burrow is difficult, since the worm quickly retreats far down into the sediment. But if successful, you will see that its soft body does not have segments or appendages like the annelid worms. An orange collar separates the white head from the long brown body. You will also notice the strong scent of a bromine-based antibiotic in the mucus excreted by the worm. The odor of the volatile bromine chemical smells to us like iodine, resulting in the other common name of this species, iodine worm.

In addition to meiofaunal copepods, a slightly larger crustacean inhabits marsh muds. The marsh tanaid, *Hargeria rapax*, is a miniature cousin of the isopods, many of which are benthic crustaceans with bodies flattened from top to bottom. These skinny little animals, only about one-eighth of an inch long, live all over the salt marsh. They feed on benthic diatoms and meiofauna, and in turn are a favorite prey of small fish and shrimp swimming into the marsh at high tide. Tanaids have a curious reproductive cycle. The females brood their eggs in a special pouch. After hatching, the baby tanaids shelter in their mother's pouch until they are big enough to survive on their own in their muddy home. But when they leave their mother, the youngsters are neuter, neither male nor female. After further molting, the tanaids develop gonads and become either adult males or females. Sometimes a female tanaid that has already produced eggs and young will, for some reason or other, reverse its sex and become a functional male. There is a huge downside to this change. Once a tanaid molts into a male, it is doomed. The mouthparts of the male are fused together, so it is unable to feed. The only duty of a male tanaid is to fertilize a female's eggs; then it starves to death.

Terrestrial insects have adapted to salty marsh soils by laying their eggs in the marsh. Fly and mosquito larvae hatch, feed, and grow to adults in marsh soils. In spring and fall, these larvae metamorphose into the annoying biting midges (no-see-ums), greenhead flies, and buzzing mosquitoes that plague the residents of the southeastern coast.

Two additional members of the mud-dwelling macrofauna are tiny air-breathing snails and little burrowing anemones. Marsh mud snails, or hydrobiid snails, only a quarter inch in size, consume the microbes on decaying *Spartina* leaves. Unlike most marine mollusks, which depend on constantly moist gills in order to breathe, the hydrobiids have tiny lungs and can survive in the relatively dry upper marsh. Although these snails are so small that they are easily overlooked, teeming throngs of hydrobiids constantly rework the marsh surface, producing nutrient-rich fecal matter.

The starlet sea anemone, *Nematostella vectensis*, a half-inch-long relative of the large anemones of rocky habitats, extrudes its tentacles from the surface of the mud to ensnare fly larvae and copepods. These miniature anemones are a food resource for grass shrimp in the marsh.

They are also useful to scientists. It turns out that the starlet anemone is easy to grow in the laboratory. Because anemones are related to jellyfish, one of the first forms of multicellular life to appear in the fossil record, molecular geneticists have been analyzing the genome of this tiny polyp for clues to the evolution of animals.

Because the abundant meiofauna and smaller macrofauna of marsh muds are easy to overlook, they are best studied by using a microscope. Larger invertebrates, too, live in the salt marsh and tidal creeks, and these include some of the most familiar wildlife of the Georgia coast. They deserve a chapter of their own.

5

Mud Dwellers of Marshes and Creeks

Most larger invertebrates, which lack the internal support structure of a bony skeleton, have an external edifice. These shells shield the soft internal tissues from damage (mainly but not always successfully). The external structure serves also as scaffolding for the attachment of muscles used for feeding and moving about. Different phyla of invertebrates have different solutions to the problem of housing delicate organs in a supporting structure. The members of Arthropoda, joint-legged creatures that encompass marine crustaceans and terrestrial insects, build a hard exoskeleton made of an organic compound, chitin. Members of Mollusca secrete a hard mineral shell in which to live. The term "shellfish" loosely includes both crabs and clams, but these animals are related only gustatorily, not by body parts or life cycles. We have met the smallest arthropods and gastropod mollusks of the marsh. Larger, much more visible mollusks and crustaceans in the marsh include the legions of black snails and fiddler crabs feeding on mudflats at low tide, and the creamy white periwinkle snails crawling up cordgrass stalks at high tide.

Mollusks

These familiar soft-bodied creatures have distinctive hard shells made of calcium carbonate absorbed from seawater. The shells are secreted by the mantle, a special organ unique to mollusks. The two main types of mollusks are the snail-like gastropods (meaning "stomach foot"), which crawl about on a slimy muscle, and the bivalves ("two shells"), which have shell halves hinged on one side. Snails, whelks, and augers are gastropods; oysters, clams, and mussels are typical bivalves. There are also shell-less mollusks, including the speedy squid and the clever octopus.

Gastropods, or snails, have a single shell that grows in a whorled spiral, and a muscular foot on which they glide about over a layer of continuously produced slime. In many gastropods, the foot is tipped with a hard operculum ("little lid"), which is shaped to fit in the opening of the shell and which the snail uses to seal itself up in time of danger or to keep from drying out. The snail's head has two eyestalks and a rasping mouth used to scrape up benthic algae or detritus or to bore into another mollusk's shell and feed on the soft tissues inside.

In addition to the minuscule air-breathing hydrobiid snails of the high marsh, three larger gastropods are abundant in salt marshes. Like their underwater brethren, two of these breathe with gills, which need to be constantly moist in order to extract oxygen from the air. One is the periwinkle snail, *Littoraria irrorata*. The white shells of *Littoraria* are hard to miss on marsh sediments in winter and on the stems of cordgrass in summer. At high tide, periwinkle snails crowd together on the tallest stems of marsh grass in order to escape the predatory crabs and fish that invade the marsh with the incoming water. This marsh snail is a close relative of the periwinkle snails of intertidal rocky shores; like the northern species, it lives on algae and other microbes rasped from surfaces.

Marsh periwinkle snails on cordgrass leaves, seeking refuge from predators swimming in on flood tide.

Also abundant, but less obvious, are the black half-inch-long mud snails *Ilyanassa obsoleta* in the lower marsh. At low tide, herds of mud snails roam the exposed creek banks and mudflats, grazing on benthic diatoms and meiofauna. Mud snails are also quick to find dead animals, their favorite food. Neither periwinkle snails nor mud snails can stray too far from the damp mud of the marsh or mudflat; they need the moisture to continually bathe their gills.

The third snail is the common marsh snail, *Melampus bidentatus*. About the size and color of a well-roasted coffee bean (the marsh species has a cousin, *Melampus coffea*, which inhabits mangrove stands and is actually called the coffee bean snail), these gastropods are found rasping over the surface of sediments in the highest parts of the salt marsh. Like the tiny hydrobiid snails in this habitat, *Melampus* is a pulmonate snail: it breathes air with lungs, not gills, so it is not tied to the tides.

In marsh creeks lives a predatory snail that goes after live animals, not detritus or carrion. The oyster drill, *Urosalpinx cinerea*, less than an inch long, has a narrow white shell with fat ribbed whorls ending in a sharp tip, and a flared opening. The oyster drill is a major predator of oysters, mussels, and barnacles. It bores a small hole in the shell of its prey by mechanical rasping, assisted by acid secretions from a special gland that dissolve the limey shell of its victim. Once the hole is made, the drill inserts its proboscis through it and feeds on the soft flesh within.

Bivalve mollusks, in contrast to roving snails, are typically sedentary animals, either attached to hard surfaces or burrowing in firm mud or sandy sediments. Bivalves feed by filtering out algae and small animals from plankton in the water. The most abundant bivalve in southern estuaries is the American oyster, *Crassostrea virginica*. Its gray shell is easily recognized by its rough, irregular shape and by the purple scar on the inside marking the attachment site of the tough muscle that holds the two shells together. Oysters are found everywhere in coastal habitats. They encrust pilings and docks and form large colonies, or oyster reefs, in salt marsh creeks.

Even so, conditions along the Georgia coast are not optimal for oysters. Suitable territory for the attachment of oyster larvae is scarce, so oysters crowd together and form tall slender shells, unlike the plump oval shells of solitary Chesapeake Bay oysters. In addition, the large tidal differences along the Georgia coast mean that oysters here spend a lot of their day out of water, shut up tightly to avoid drying out, while oysters in other estuaries are able to feed continually underwater. The resulting small slender oysters, which eke out a precarious existence in the creek beds, are known as "clutch oysters."

Oysters are prime food, nutritious and easy to collect. On Sapelo Island, as on other sea islands, there are many old oyster-shell middens left by previous inhabitants. When I lived on Sapelo, we often had oyster roasts under the oak trees and made some small shell heaps of our own.

This delectable shellfish was the basis of a great fishery along the Atlantic and Gulf Coasts of North America from the first settlement by European colonists up until World War II. (For an amusing and fact-filled account of the nineteenth-century U.S. oyster industry, try

Oyster reefs built up in marsh creeks offer shelter to many marsh animals and provide tasty meals for raccoons and people.

Mark Kurlansky's best seller *The Big Oyster: History on the Half Shell*.) Despite the suboptimal growing conditions in muddy salt marsh estuaries, oystering was a main source of income for many families along the southeastern Atlantic Coast, more so in South Carolina than in Georgia. Collecting, shucking, and canning oysters are hands-on jobs that required cheap, plentiful labor. Harold Coffin had a small canning operation for shrimp and oysters on Sapelo Island in the 1920s. Oyster production in southeastern estuaries peaked during the years between the world wars. Although there was a bump up in oyster production after introduction of mechanical shuckers in the mid-1940s, oyster fisheries declined sharply after 1950 everywhere along the Atlantic and Gulf Coasts. There were several reasons for the decrease. Coastal development led to the filling, polluting, and silting of prime oyster habitat; shucked oyster shells were not returned to estuaries for the resettlement and growth of oyster larvae; and better job opportunities in the booming postwar economy meant fewer people willing to work for low pay in oyster factories.

Adding to these trends, in the 1950s a major shellfish disease, dermo, was discovered infecting oysters in Chesapeake Bay, Virginia. The dermo disease is caused by a flagellated protist that is ingested by an oyster, whereupon the single-celled invader insinuates itself into the oyster's blood cells and starts reproducing. The parasites fill up the blood cell until it bursts, releasing more infective protists throughout the oyster's body. The oyster eventually fills with protists. The oyster becomes sick, stops growing, and may die. At death, the shell of the oyster gapes open and the oyster's body disintegrates, releasing millions of the dermo parasites to find other oysters to infect. No wonder that this disease is so highly contagious; after the original outbreak, it soon spread to oyster beds in estuaries from Virginia to Louisiana. A study of the disease in oysters in the Duplin River along Sapelo Island found 90 percent to 100 percent of the bivalves infected in 1999 and 2000. Most of the oysters don't die, but their growth and food quality are affected by the parasite.

In colonial times, settlers along the Atlantic Coast used oyster shells when constructing walls of tabby stone, a kind of cement made from lime, sand, water, and crushed shells left over from oyster harvests. Tabby ruins from a nineteenth-century plantation built by a Danish sea merchant still stand on Sapelo Island. The name of this old plantation, Chocolate, may be a corruption of the name of a local Native American village called Chucalate.

Oyster reefs, built up as these bivalves attach to one another to gain a shell-hold in the muddy bottom of the estuary, create new habitats for multitudes of other creatures. Oyster reefs are protective "cities" for animals that could not easily survive without such a refuge from the sweep of tidal currents. Some species have adapted so well to their shelly haven that they are found nowhere else. The yellow boring sponges that infest oyster shells are preyed upon by a yellow-orange nudibranch (shell-less mollusks that look something like exotic slugs). Several types of bristle worms also bore into oyster shells. Small mud crabs, shrimp, and brittle stars live among the oysters, which can become encrusted with algae, sponges, and bryozoans. Gobies and other small fish feed on the abundant denizens of the oyster reef when the tide is high. Often the reefs stretch completely across creeks, damming the channels. When the tide retreats, small pools remain behind the oyster reef dams, giving shelter to shrimp, crabs, and killifish that had foraged in the marsh on the flood tide.

Up in the intertidal cordgrass marsh one finds another bivalve, the ribbed mussel, *Geukensia demissa*. This mussel has a dark brown, nearly black shell, up to four inches long, with defined vertical ridges, or ribs. The mollusk inside is tinged yellowish brown. Ribbed mussels are common in *Spartina* marsh plains, embedded in the mud in small groups attached to the grass stems by tough byssal threads. (The byssus is the gland that secretes these sturdy filaments.) They are sometimes found lodged in crevices in creek-bank oyster reefs. Juvenile ribbed mussels not carried by the tide to a proper home in the marsh may instead settle out in fouling communities on pilings. Mussels filter plankton from tidal water flooding the marsh, and add nutrient-rich undigested matter to the marsh surface from their abundant feces. Excretions from a clump of mussels can create a mound of feces and spit-out mud several inches higher than the surrounding marsh

Several tightly closed ribbed mussels poke up from the mud in a cordgrass marsh. When tides flood the marsh, the mussels open up to filter plankton from the water and deposit mounds of fecal material on the marsh surface.

mud. Over time, combined deposits from the mud-dwelling mollusks build up the surface height of the marsh.

Because these mussels live in sediments oozing with noxious sulfide, their flesh also stinks, and only adventurous foodies would want to try them. Marsh predators aren't as picky. Blue crabs swimming among the cordgrass at flood tide can crush the top of the mussel shell and pick out the flesh. Raccoons find them tasty. Clapper rails likewise feed on ribbed mussels, but not without a cost. Investigators have observed clapper rails that must have stuck a toe into a live mussel while trying to open the shell. The result was a rail stalking around with a mussel tightly clamped to a foot, or even with a missing toe.

The tidal creek bottoms are an additional bivalve habitat. One day as I was watching tidal water draining from a marsh creek, I was startled to see a jet of water spurt over a foot into the air from a small hole in the creek bottom. A few seconds later, another long squirt was ejected from a nearby hole, and over the next few minutes there were a dozen more jets all across the creek. Since clams are well-known squirters, I figured it must be a colony of hard-shell clams. Hard-shell, or northern quahog, clams, *Mercenaria mercenaria*, live in sandy sediments where water currents are strong: lower-beach sandflats and large tidal creeks. Their thick white oval shells are up to four inches across and, like oyster shells, have a purple scar on the smooth, pearly inside. These clams can be used to make tasty homemade chowder. Our family went hunting for quahogs by feeling along suitable creek bottoms with our bare feet. When our toes felt something hard in the muck, it was likely a clam.

Another clam species has staked out living space in the opposite direction, in the high marsh. The marsh clam, *Polymesoda caroliniana*, less than two inches long, resembles an immature hard-shell clam. Marsh clams, which prefer less saline conditions than other estuarine bivalves do, are found spottily along the borders of salt marshes where there is some freshwater drainage.

Crustaceans

If you turn over a mass of cordgrass stems washed up on the beach, bunches of flea-sized bugs will start hopping madly about. Poke around a rock-filled bulkhead at the shore, and you might see what appear to be giant dark gray roaches scurrying away. These are examples of two types of coastal crustaceans, each with seven pairs of legs. The beach fleas are amphipods, their bodies flattened from side to side. The sea roaches are isopods, flattened from top to bottom.

Amphipods are generally more obvious than isopods in coastal habitats, because of their overwhelming abundance. Marsh scuds, *Gammarus palustris* and *Gammarus mucronatus*, usually no more than half an inch long and translucent green to sandy brown in color, live in great numbers in masses of dead plant stems and leaves in the salt marsh. These amphipods also live among the algae, hydroids, and bryozoans of fouling communities, along with another very distinctive amphipod, the skeleton shrimp. The latter creature, in the genus *Caprella*, resembles a tiny praying mantis. Skeleton shrimp sit in wait for other small animals to venture past, then leap out to seize their prey with long front legs.

The common isopod of intertidal sediments is the inch-long *Cyathura polita*. It is unusual to encounter these isopods unless one is actively looking for them. One who did seek out these little animals was my undergraduate mentor at Emory University, William Burbanck, who first introduced me to the Georgia estuary. A species in this isopod group, *Cyathura burbancki*, is named in honor of his research.

Sea roaches are the best-known marine isopods; they infest docks, seawalls, and rocky shores all over the world. The southeastern species is *Ligia exotica*, the exotic sea roach. North of Cape Cod, a different species, *Ligia oceanica*, occurs. These large isopods have an uncanny resemblance to cockroaches in size, color, and behavior. Sea roaches are just as quick and secretive as their terrestrial namesakes, zipping into tight crannies at the least disturbance. Sea roaches can be spotted wherever rock seawalls have been placed to protect shorelines—for example, along the beach by the fishing pier at the southern tip of St. Simons Island or along stretches of beach on Jekyll Island where erosion has been a problem.

Smaller, pill-bug-like isopods, less than half an inch long, are found around oyster reefs and in fouling communities. Some species of isopods live as parasites on fish gills. The most economically notorious isopods are gribbles, species of *Limnoria*. These crustaceans riddle pilings as they feed on marine fungus growing in the wood. Gribbles do their worst just above the low-tide line. As the tunneling activity of gribbles erodes the wood at this level, old pilings start to look as though a giant beaver had been gnawing them.

Decapod ("ten-footed") crustaceans, including crabs and shrimp, typically have a body divided into a head with eyestalks, a thorax covered with a hard carapace, a softer abdomen, and five pairs of jointed walking legs, some of which end in claws. A signal feature of decapods is the set of greatly enlarged claws on the first pair of walking legs. Like all arthropods, decapods must shed, or molt, their chitin exoskeletons as they grow larger. Decapods, with the exception of penaeid shrimp, brood their eggs as a mass, known as a sponge, on special abdominal appendages, the pleopods.

Two smaller shrimp species are permanent residents of the estuary. Grass shrimp, *Paleomonetes pugio*, are the most common. Adult grass shrimp are one to two inches long and have transparent bodies. They can be distinguished from young penaeid shrimp by the two pairs of little claws that grass shrimp have on their walking legs; penaeids have three sets of claws. Multitudes of grass shrimp congregate in tidal creeks that drain the salt marshes. During summer at low tide, pools in the marsh creeks churn with grass shrimp frantically trying to escape fish that swim into the marshes to feast on them.

The other ubiquitous shrimp of the estuary is the snapping shrimp, *Alpheus heterochaelis*. Looking more like a tiny lobster than a shrimp, this species, which reaches about two inches in length, is identified by its lack of a long rostrum (a beak-like structure extending past the eyes) and by its one large claw. This claw, almost half the length of the body, is an amazing weapon—the shrimp equivalent of a Saturday night special. The claw has a tooth that fits tightly into a socket. The tooth can be cocked just like the hammer of a gun; when it is suddenly popped into the socket, a powerful snapping sound is made. *Alpheus* lives in a subtidal burrow, lying in wait at the entrance. When its long antennae, which protrude from the burrow's entrance, detect a small fish or other likely prey nearby, the snapping shrimp sneaks out and lets its victim have it. The sudden shock wave from the claw's snap stuns the prey, which the shrimp then finishes off and drags back down its hole for a leisurely meal. At low tide, grass shrimp and young killifish massed in tidal creek pools behind oyster reefs are easy pickings for the pistol-packing shrimp, and a staccato of sharp pops often resounds over the marsh.

The burrowing mud shrimp, *Upogebia affinis*, which is related to the beach-dwelling ghost shrimp, prefers calmer, estuarine habitats to ocean surf. In compact clay sediments and muddy sands, mud shrimp build elaborate complexes of branching tunnels, half an inch in diameter and up to three feet deep, in salt marsh creek banks and down to bottom depths of thirty feet in the open estuary. Often, several mud shrimp will live together in one interconnecting burrow system with a number of constricted openings to the surface. Like the ghost shrimp, *Upogebia* feeds by straining plankton from water forced through the burrow by the pumping action of its appendages.

Stomatopods, commonly known as mantis shrimp, are stumpy, flattened crustaceans that can grow up to ten inches long. Mantis shrimp are equipped with a pair of claws that are highly dangerous weapons. The outer part of the claw is a smooth, razor-like blade that fits into a groove along the inner section of the claw, just like a jackknife. When the shrimp is within striking distance of intended prey, the knife blade part of the claw whips out incredibly fast to deliver a lethal slash. Two species of these creatures in Georgia estuaries are *Squilla empusa* and *Gibbesia neglecta*. Both live in subtidal burrows and are rarely seen except in trawls; if you encounter a mantis shrimp, use caution in order to avoid getting a nasty cut from the jackknife claws.

Unlike shrimp and anomuran ("unusually tailed") crabs, such as hermits, the bodies of true crabs, or brachyurans ("short-tailed"), are mostly a hard carapace, with a much-reduced abdomen. Crabs, like shrimp, are omnivorous, which means they will eat anything that doesn't eat them first. Crabs that live in the intertidal zone have modified gills that allow them to breathe in air; they need water only to keep their gills wet. But these crabs are still tied to the sea; they have to release their larvae into the ocean or estuary when it is time for the brooded eggs to hatch. The larval crabs go through two stages in the plankton. First, they are shrimp-like zoea, often with long spines on their heads; then they metamorphose into tiny crab-like creatures with long abdomens, the megalopa stage, after which they leave the plankton to settle onto the marsh mud or beach sand.

The most numerous crabs of the marshes are creatures more of land than of sea. Fiddler crabs tunnel into the salt marsh, emerging during low tide to feed on the mud surface. During summer, fiddlers swarm in vast throngs over the marsh and the sandflats. I was always reminded of herds of antelope on the African plains.

Three species of fiddler crabs are common in Georgia estuaries; each is specially adapted to feeding on different types of sediment. The dark mud fiddler, *Uca pugnax*, is found over most of the *Spartina* marsh, from muddy creek banks to the firm mud of the marsh plain. The mud fiddler has a sieve of hairs ending in little spoons on its mandibles, or mouth appendages, which allows it to scrape algae and other

Fiddler crabs scavenging in cordgrass marsh at low tide.

microbes from sandy mud. The lighter-colored sand fiddler, *Uca pugilator*, is confined to the upper marsh and around lagoons at the beach. Additional hairs on its mouth appendages enable it to efficiently sort out food particles from sand grains. The red-jointed fiddler, *Uca minax*, lives in fresher-water tidal habitats, along the banks of coastal rivers, in a burrow it digs in brackish stands of black needlerush. It has only a few hairs on its mandibles, all that it needs to process silty mud.

These marsh crabs spend a good deal of time out in the air, scavenging on the surface of the marsh during low tide. Like all crabs, the fiddlers have gills that need to be kept moist so that they can breathe. Fiddler crabs have a special enclosed gill chamber in which they trap water to keep their gills moist. While the crabs scurry about in the hot sun, the water in the gill chamber slowly evaporates. When a crab's gills get too dry, it retreats to its burrow to replenish the gill chamber with water accumulated at the bottom of the hole. Then, it can come out again to scrape algae, meiofauna, and detritus from the marsh surface. Fiddlers are messy eaters. The feeding claws transfer globs of mud to the mouth, and mandibles extract tasty tidbits. Rejected material dribbles down the front of the crab. Every now and then, the crab uses its feeding claw to scrape off the muddy drool, which it deposits as a small ball of mud. The marsh is littered with little gray spitballs after a crowd of fiddlers has passed by.

Male fiddler crabs have one large claw, or "fiddle," which they wave about to defend their territory, engage in courtship displays, or fend off any human who tries to catch them. The large fiddle claw is something of a liability for the male crabs. They can't use it to feed, leaving them with only one small claw to move juicy tidbits to their mouths. Female fiddlers have two feeding claws, so they can engage in double-fisted dining. Fiddlers are aptly named, since males use their big claws to make music to attract females or to warn off rivals. By scraping the fiddle against their hard carapaces or against the ground, males create different sounds to signal their intentions. Male fiddlers also "blush" when courting. During the day, mud fiddlers are usually dark colored, blending in with the marsh surface. But when flirting with a prospective mate, a male will lighten up his color, or blanch, to attract attention.

A male fiddler crab, *Uca thayeri*, caught in a salt marsh in Camden County, Georgia. The crab's "fiddle" claw is aggressively pinching a finger. This warm-water species, common in Florida mangroves, has recently been spotted in Georgia, perhaps as a result of climate change. The large claw is orange, unlike those of other marsh fiddlers. Photo provided with permission by Dr. Clay Montague.

If the courtship is successful, the male will take his ladylove down into the burrow he has tunneled between the cordgrass roots and fertilize her eggs. If mating occurs during an evening ebb tide, the crabs may feel safe enough to have their tryst topside, under the moon and stars. Because fiddler crabs spend much of their time out of water, and have to mate quickly, they can't afford to be locked together until the female molts, as blue crabs do. The male transfers his sperm to the female even though her shell is hard.

The gravid female retreats underground, bathing her egg mass, the sponge, in the water at the bottom of a burrow until the youngsters are ready to hatch. Then she migrates to the edge of the marsh during a high tide and waves her abdomen up and down, releasing the little fiddler larvae into the sea. They will grow on algae in the plankton until, after a month or so, they undergo a final transformation into miniature fiddlers, less than an eighth of an inch long. The little crabs settle into the salt marsh. When they are just a bit larger, they begin digging their own tiny burrows in the mud. After a year, they will mature into adults, ready to breed. But life is rough in the marsh, where there are many crab hunters. Most fiddlers do not survive more than two years.

Two types of squareback crabs live in the salt marsh. About the size of fiddler crabs, squareback crabs have a flat, rectangular carapace. The wharf crab, *Armases cinereum*, is light in color and usually found around docks or at the upper edge of the high marsh, sometimes fairly far from water. The second species, *Sesarma reticulatum*, is darker and common in the muddy parts of salt marshes, where it lives in communal burrows and grazes on cordgrass leaves, often cutting tender shoots off at the base.

Other bottom-dwelling crabs are found in wetter habitats in intertidal sediments. The abundant marsh mud crab, *Eurytium limosum*, which has a dark chunky body up to one and a half inches wide, is found only in southeastern estuaries. During low tide, the mud crab shelters in water-filled burrows in the salt marsh, emerging at flood tide to feed on baby killifish, which it crushes with its massive claws. Two smaller relatives of *Eurytium* are the black-fingered mud crabs, *Panopeus herbstii* and *Eurypanopeus depressus*. These are similar in shape to the marsh mud crab, but have dark-tipped claws and are more commonly found on subtidal mudflats and around oyster reefs, where they devour newly settled baby oysters.

Besides being entertaining to watch, these small marsh crabs are a vital part of the diet of estuarine fish, birds, and mammals. At high tide, marsh crabs retreat to their burrows to escape predators, such as blue crabs and fish, swimming over the marsh in search of food.

A few species of terrestrial arthropods have also managed to find a home in the salt marsh. Insects and spiders, distant relatives of the crabs scurrying over the marsh mud, have their own roles to play (and people to annoy) in Georgia's coastal ecosystems.

Creepy Crawlies   *Insects and Spiders*

One late summer day while walking along the edge of a *Spartina* marsh, I was suddenly enveloped by a blizzard of tiny white floating specks. At first I thought they might be the fluffy seeds of the bordering groundsel bushes. Then I noticed that some of the "seeds" had landed on my arms, and had wings. They were minuscule flying insects, most likely plant hoppers, which spend the summer sucking plant juices from cordgrass leaves and stems. I was grateful they weren't trying to suck my sap, which is more than could be said for the mosquitoes and no-see-um gnats also hovering around me.

Coastal Georgia is definitely an entomologist's paradise. Hundreds of species of insects and their close relatives, the spiders, live in southeastern marshes and beach habitats. Biting midges, flies, and mosquitoes swarm along the coast. Residents on Sapelo Island divided the year not by weather, but by bugs. Spring and fall were biting gnat seasons, summer was biting fly season, and mosquitoes were around all the time except in the cold of winter. Insects and spiders have even found special niches in the salt marshes, which are rather inhospitable to most terrestrial animals.

The salt marsh plant hopper, *Prokelisia marginata*, is one of only a very few animals able to feed on living cordgrass. Occurring in enormous multitudes during the warm months, marsh hoppers enjoy a liquid diet of plant juices, extracted through a hollow proboscis stuck into cordgrass tissues. In turn, the little insects are food for marsh spiders and birds. At high tide, when plant hoppers crowd the emergent leaves of *Spartina*, killifish often are seen leaping up out of the water to feast on the concentrated swarm.

The salt marsh grasshopper, *Orchelimum fidicinium*, eats live cordgrass by rasping tender leaf shoots. A rather ordinary-looking small green grasshopper, this species is the only one adapted to life in the *Spartina* marsh. Salt marsh grasshoppers are hunted by cattle egrets stalking through the grass at low tide. During high tides, seagulls have been observed picking off grasshoppers massed together on the tips of cordgrass leaves poking up from the water. Even though plant hoppers and grasshoppers are abundant, these insects consume less than 10 percent of the annual production of cordgrass.

An ant species, *Crematogaster laeviuscula*, a member of a group known as "acrobat ants," has also found a home in the marsh. Colonies of this small brown ant live inside hollow stems of *Spartina* and feed on sap oozing from plant tissues. At high tide, one of the ants blocks the entrance of the nest with its specially adapted head to prevent the colony from flooding.

On foggy mornings, the abundance of spiders in the marsh is attested to by the multitude of webs, highlighted by dew droplets clinging to the strands spun between stems and leaves of cordgrass. A technician who worked at the University of Georgia Marine Institute for several years was fascinated by the marsh spiders. Studying them became such an obsession that he literally spent all his spare time at it. Naturally, he became known as the resident "spider man." After attempting to gather samples of marsh spiders by doing traditional net sweeps, he decided that to be really thorough he had to harvest all the marsh grass within a measured plot, bring the bagged grass back to the lab, and carefully inspect each leaf for spiders. This he did every night for a couple of years, in a tiny cubicle, with his radio tuned to the classical music station in Jacksonville, Florida (this was before there was public radio in Savannah). After becoming a graduate student, he got funds to sample the spiders more efficiently. He would set up a big plastic-enclosed frame over a marsh plot and then vacuum the buggers right off the cordgrass with a huge industrial vacuum rigged up in a backpack. Among the numerous arachnids he found, the most common were two spider species, *Grammonota trivittata* and *Clubiona littoralis*, and a species of pseudoscorpion in the genus *Paisochelifer*.

Spiderweb outlined by fog droplets at the edge of a cordgrass marsh. Spiders catch marsh planthoppers, flies, and mosquitoes in their webs, and in turn are food for marsh birds.

He discovered that the marsh spiders mainly caught plant hoppers in their webs, and that spiders were a favorite food for marsh wrens and sparrows.

The most infamous, and familiar, coastal insects are the biting pests: midges (or gnats), flies, and mosquitoes. The males of these insects are innocuous, living on plant sap during their brief lives. It is the females that bite to obtain a blood meal for the complete development of their eggs. Tiny biting midges, species of *Culicoides*, known as sand gnats or no-see-ums (they really are hard to see), swarm around docks, beaches, and marshes in the spring and fall. Midge larvae live in salt marsh soils or in beach sand. A more painful biter, however, is the greenhead fly, *Tabanus nigrovittatus*. A large fly with bottle-green eyes, the greenhead hatches out of salt marsh soils in late spring, about the time when the no-see-ums die down. A smaller hatching of greenheads happens in early fall. The tabanid flies take a chunk of flesh when they bite. Deer flies, species of *Chrysops*, small black flies resembling houseflies with striped wings, also have a painful bite. When breeding, deer fly larvae prefer less saline muds than the greenheads, so are most common in the more brackish upper marsh and riverine habitats.

Of all the coastal insects, the most research has been done on tidal marsh mosquitoes. ("Mosquito" is a Spanish word meaning "little fly.") In the northeastern United States, salt marshes are pocked with shallow pools of water that are perfect for the larval development of saltwater-breeding mosquitoes. Early attempts to reduce the coastal mosquito population by gridding the marsh with ditches that would drain the pools largely failed, mainly because the ditches were not well connected to standing water on the marsh. A more successful management plan has been to dig ditches directly from tidal pools to the nearest creek, thus effectively draining the water from the marsh. Such surface pools are rare in southern salt marshes. Here, the most prevalent marsh mosquitoes are species of the genus *Aedes*, which, unlike their northern relatives, deposit single eggs on moist plant litter and mud. After they hatch, the mosquito larvae find small water-filled depressions on the marsh surface, where they feed on algae and microbes growing on plant detritus. The two common salt marsh mosquitoes in Georgia are *Aedes taeniorhynchus* and *Aedes sollicitans*, the latter being the more aggressive biter of humans.

Insects and spiders are also abundant in beach dunes; most species there have a light, mottled coloration that allows them to blend in with their sandy environment. The tan color of the dune wolf spider helps it hide from predators as well as ambush its prey. The dune spider lives in firm sand in a hole with a trapdoor, from which it leaps out to seize passing insects. Like other wolf spiders, the dune spiders are most active at night. One can easily spot prowling spiders in the dark by holding a flashlight at eye level and shining it over the dunes. The compound eyes of the spiders reflect the light like tiny jewels in the sand.

Ant lions (species of *Glenurus*) are another common predator in the dunes. These larvae of winged insects that resemble delicate damselflies create conical depressions in the sand that are traps for unwary insects. The sandy pits give the ant lions their other common name: doodlebugs. The unfortunate ant or beetle that happens into the ant lion's trap slithers on falling sand grains to the bottom of the depression and into the waiting jaws of the larva buried just under the sand at the bottom of the cone. One such unlucky bug might be a brightly striped tiger beetle, of the genus *Cicindela*, a roving dune hunter of insects and amphipods.

Several other insects are spectacularly visible in coastal Georgia in late summer and fall. Coastal residents are all too familiar with lovebugs, *Plecia neartica*, which, oblivious of traffic, fly in dense swarms over highways, necessitating the frequent washing of cars. This fly emigrated from South America in the middle of the twentieth century. The swarming stage in late summer is primarily devoted to mating; male and female lovebugs remain coupled for five to seven days. The female lays a batch of eggs in an open field, where the young larvae hatch out and move together in a squirming mass, feeding on dead plant detritus. The flies that hatch in the fall overwinter on the ground, where they make a favorite snack for migrating robins. In May, the surviving larvae pupate and form a smaller spring swarm of lovebugs. The larvae hatched from eggs laid in early summer grow fast and have a higher rate of survival; thus the much greater numbers of pupating larvae in late summer produce the large September swarms of these bugs.

In fall, sitting at the edge of the beach dunes on a clear, calm day, I have been treated to a constant fluttering of yellow and orange butterflies, all moving south. These are the cloudless sulfur butterfly, *Phoebis sennae*, and the Gulf fritillary butterfly, *Agraulis vanillae*. During summer, these butterflies are found all over the Southeast and Midwest, as far north as Virginia and Kansas. They travel south for the winter, heading for central Florida. Unlike the most famous butterfly migrator, the monarch, which catches high air currents en route to Mexico, the cloudless sulfur and Gulf fritillary are boundary-layer migrants. They travel close to the ground in order to avoid winds that might blow them off course. Sensibly, they will not fly over large stretches of open water, so when they arrive at a coast they begin following the line of dunes to their winter destination.

7

Marsh Life *Scales*

At high tide, most of the crabs, snails, and insects scurry into their burrows or swarm up cordgrass leaves jutting above the rising water for protection from blue crabs and small fish swimming in from the tidal creeks. At low tide, the marsh inhabitants are not safe either, since land-dwelling island denizens stalk the exposed mud for a banquet of insects, crustaceans, or mollusks. These scaled, scaly, feathered, or furry beasts are vertebrates, animals with a backbone and an internal skeleton. Most are not permanent citizens of the salt marsh, only immigrants looking for a good meal.

The Scaled: Fish

If you wade out into a salt marsh at high tide, you are likely to notice schools of minnow-like fish darting about the cordgrass stalks. These small, broad-headed killifish ("killi-" derives from a Dutch word meaning "small creek") are the most abundant marsh vertebrates. They prey on small invertebrates and in turn are eaten by wading birds and larger fish. The greenish mummichog, *Fundulus heteroclitus*, is the most widely distributed. The common name of this killifish derives from a Narragansett word meaning "going about in crowds," referring to the fishes' habit of schooling in the flooded marshes. These killifish will eat almost any small animal that moves in the marsh. Their staple diet consists of the little crustaceans and bristle worms that live in the marsh muds. But these fish are always on the lookout for other prey. While wading into a marsh during a high summer tide, I have seen mummichogs leaping up out of the water to dislodge plant hoppers and spiders crowded on emergent cordgrass leaves. The fallen bugs were then avidly snatched from the water surface by the killifish.

At low tide from spring to fall, tiny larval mummichogs dart about in small puddles left on the surface of the marsh by the retreating water. Often these pools of water are at the entrances of fiddler crab burrows. These depressions would seem an inhospitable place for the baby fish to wait for the tide's return. Under the hot summer sun, the temperature in the pools can exceed 100°F, and at times the refuges completely dry up. But studies have shown that killifish hatchlings have a better chance of surviving by remaining in the marsh. During low tide, hungry hordes of grass shrimp in the tidal creeks would make short work of larval fish.

Mummichog, a common salt marsh killifish. Photo from Wikipedia Commons, taken at St. Michaels, Chesapeake Bay, in 2006 by Brian Gratwicke. Used with permission.

To ensure that their babies will be protected in the upper marsh, mummichogs spawn in short *Spartina* marsh on the highest monthly tides. Eggs are laid in places that will stay moist during low tide; favorite sites are empty ribbed mussel shells and the inside surface of leaves at the base of cordgrass shoots. The eggs mature in a few weeks, but the tiny larvae will not hatch until the eggs are immersed in water on another high flood tide. Then they emerge very quickly. Larval mummichogs spend their first days in small shallow depressions that pocket the surface of the marsh. There, besides avoiding predatory grass shrimp, they can continuously feed on harpacticoid copepods in marsh soils; and they grow quickly. When the young fish reach about half an inch long, they are strong enough to escape shrimps' claws and can safely retreat to the marsh creeks at low tide. As full-size, adult killifish, mummichogs are able to exact their revenge by feasting on the grass shrimp that threatened them as larvae.

Other killifish species are common in the marshes and estuary. The spotfin killifish, *Fundulus luciae*, prefers higher marsh habitats; the brownish marsh killifish, *Fundulus confluentus*, lives in brackish to freshwater creeks and pools. The striped killifish, *Fundulus majalis*, swims in beach surf. Female striped killifish have several irregular black stripes overlying vertical dark stripes on their sides; males of this species have only the vertical stripes and, when breeding, a yellow tail.

A college instructor on a day trip to Sapelo once asked me where he could find sheepshead minnows. The sheepshead minnow, *Cyprinodon variegatus*, lives in both salt marsh tidal creeks and in low-salinity pools and ditches. Although about the size of a mummichog, this killifish has a broader body and shorter tail. Normally drab, the male sheepshead assumes iridescent rainbow colors when breeding. I really hadn't collected any, but we tried an interdune slough at Cabretta Beach, which was nearly fresh because of recent rains. The first sweep of his net in the stagnant, weed-choked water yielded a large beautiful male *Cyprinodon* in full breeding color. I was astonished, but could not dispel the teacher's impression that I knew everything about collecting on the island.

Another little fish, the sailfin molly, *Poecilia latipinna*, is found in the same habitats as the sheepshead minnow. This small, live-bearing, guppy-like fish can be distinguished from the egg-laying killifish by the molly's large dorsal fin and the parallel lines of dark spots along the length of the body. Both the sheepshead minnow and sailfin molly adapt well to freshwater aquaria.

The most comical small fish of the tidal marsh is the naked goby, *Gobiosoma bosc*. The naked goby and related species are small mud-colored fish with elongated bodies and squarish heads sporting big, googly eyes. Gobies are bottom dwellers; look for them around oyster reefs and in marsh creek tidal pools. The gobies' pelvic fins are modified to form a sucking disc that lets them easily stick to surfaces. This adaptation helps them shelter within empty oyster shells while swift tidal currents flow past.

The Scaly: Reptiles

Not many reptiles manage to live in saltwater habitats, but those that do flourish. Along the southeastern U.S. coast, the three reptiles best adapted to coastal habitats are the American alligator and the diamondback terrapin in the salt marshes, and the loggerhead turtle in offshore waters. Other reptiles commonly encountered on the barrier islands are lizards, the most abundant being the green anole, and snakes, two of which, the cottonmouth and the diamondback rattlesnake, frequent the edges of the salt marshes and the beach dunes.

The iconic reptile of southern swamps and marshes is the American alligator, *Alligator mississippiensis*. The lineage of alligators and crocodiles is extremely old; these animals are essentially unchanged since the great age of dinosaurs. They have gotten a bit smaller in size. During the Cretaceous period, a giant crocodile grew to forty feet long, as long as a school bus, and weighed ten tons. American alligators mainly live in freshwater. They don't have adaptations that allow them to excrete excess salt, as sea turtles do. But alligators can survive for a time in saltwater, and are often seen cruising in estuarine sounds and far up tidal creeks in search of fish during the summer and fall. Sometimes an alligator will become disoriented, swim out the mouth of a sound, and crawl up onto the shore. One should give any alligator seen on a beach a very wide berth. Even if it looks dead, it likely isn't.

During the southeastern drought of 1986, many of the small ponds and "gator holes" on the sea islands dried up. Alligators were forced to congregate in larger ponds. At the peak of the dry spell, there were as many as forty alligators in the freshwater pond by the Marine Institute, from a ten-foot-long adult bull to dozens of two-foot-long juveniles. The gators were hard-pressed for food, so every evening they would wander into the nearest salt marshes to search for fish and birds. To do that, they had to cross the sandy road running out behind the laboratory. One dark night as I was driving to the lab to check on an experiment, I noticed what seemed to be a new tree stump at the edge of the lane. All of a sudden the stump reared up on four legs and lurched right under my car, which bumped and thunked over the obstruction. When I managed to stop and look back, I saw a five-foot-long alligator in the middle of the road, a bit dazed but otherwise unharmed as it continued toward the marsh.

The Carolina diamondback terrapin, *Malaclemys terrapin*, is a small turtle that lives year-round in salt marshes and tidal creeks. The females are somewhat larger, six to eight inches long, than their mates, which at their manliest are only about five inches long. Clay Montague, who spent years roaming around Georgia estuaries, said that he once saw a convention of about a hundred little male terrapins swimming in a tidal creek, perhaps all bent on winning a female's favors. The terrapins' main foods are the abundant periwinkle snails and fiddler crabs that live in the marsh, as well as tiny clams in the creeks. Richard Heard, a venerable coastal ecologist who did research in the marshes around Sapelo Island in the late 1960s, told me of watching terrapins carefully scrape off the legs of fiddler crabs they had caught, to avoid the prickly claws going down their gullets.

In early summer, female terrapins deposit a clutch of five to ten eggs in nests dug at the edge of the marsh, just above the highest tide line. In August and September, the baby turtles hatch out and head for the marsh. In the early twentieth century, large numbers of terrapins were caught in southeastern marshes and sold to northeastern restaurants as a gourmet delicacy.

These days, terrapins have legal protection, and their chief predators are not human. Their eggs are great food for snakes, which unhinge their jaws to ingest eggs whole, break the tough covering with strong stomach muscles, and then digest the contents. Clay Montague told me that he has observed small red-black-and-white-striped scarlet snakes, *Cemophora coccinea*, which resemble coral snakes but are not venomous, raiding terrapin nests. Scarlet snakes are well known egg eaters. Larger snakes, including rat snakes (*Pantherophis* sp.), eastern king snakes (*Lampropeltis getula*), and black racers (*Coluber constrictor*), also go after turtle eggs. Nests missed by snakes are still vulnerable to wily raccoons (*Procyon lotor*), which dig up freshly laid eggs from the nests and also pounce upon newly hatched turtles.

A lizard common all over the Georgia coast, the green anole, *Anolis carolinensis*, hunts insects in bushes along the edge of the marsh. Chameleon-like, anoles can change the color of their skin from green to brown, depending on whether they are resting on a leaf or a branch. During the spring breeding season, male anoles puff out a bright red throat patch as a territorial display. Anoles are alert and quick, thus hard to catch. Once I sneaked up on a male sitting on a palm frond busily advertising his whereabouts to rival suitors and potential mates. I grabbed the little lizard before he saw me, but I let him go just as quickly. Despite their small size, adult anoles are capable of giving a good bite on a finger. After that, I confined my lizard catching to tiny baby anoles, just one or two inches long, with correspondingly small jaws, which seem to be everywhere a few weeks after the anoles' mating season has run its course.

Two poisonous snakes live around salt marshes and ponds in the back dunes. The eastern cottonmouth, *Agkistrodon piscivorus*, likes moist habitats, including the edge of the marsh, where it often hunts for small fish, frogs, and mammals. The venom of this snake produces a nasty wound because of the necrosis, or death, of tissues around the bite. Our border collie, Yofi, had a run-in with a cottonmouth. She was bitten in the pit of her foreleg during one of her forays into the marsh, and fortunately managed to crawl to the road to the lab before collapsing. When a neighbor found her, her front leg was enormously swollen and oozing blood and pus. We had no idea what had happened, but when we got Yofi to a vet on the mainland, he quickly discovered the wound was from a snakebite. Yofi was treated with antibiotics to minimize infection, but her leg took weeks to completely heal. After that, we all became much more careful when venturing into marshy places.

The eastern diamondback rattlesnake, *Crotalus adamanteus*, is even more common on the Georgia coast. This large venomous snake is easily recognized by the dark brown diamond pattern on its back, not to mention the rattles on the tail. Island rattlers can grow up to eight feet in length; since they are thick bodied, a large rattlesnake is formidable. Diamondbacks are expert swimmers and have occasionally been spotted in the ocean several miles from land. These snakes are most common in beach dunes and around marsh edges, where rodents and small birds are plentiful. Since rattlers prefer not to attack creatures much bigger than themselves, cases of rattlesnake bites are fortunately rare, occurring only when a snake is surprised or cornered. In one instance, a graduate student living on Sapelo was jogging down an overgrown track into the dunes when he noticed something dragging on one of his shoes. It was a small rattler that he had apparently stepped on, which had struck at his foot and gotten its fangs stuck in the heel of his sneaker. He kept on running until the snake worked its way free and dropped off.

While living on Sapelo, my family and I didn't see many rattlesnakes, but then we didn't go out of our way looking for them either. The closest I came to one was at the front door of a neighbor's house. I was bringing them their Sunday newspaper just after they had returned from an off-island trip. As I turned to leave, I was shocked to see a tremendous rattler (it was about four feet long) stretched out on the walkway just inches (well, maybe thirty-six inches) from my feet. I yelled and burst through the open door of the house, and the snake, upset by all the commotion, took off the other way. Our other experiences with rattlers were gustatory. When nuisance snakes were killed on the island, their meat didn't go to waste. Either southern fried or barbecued, the cooked flesh is a lot like the breast meat of chicken, except for the funny little rib bones.

8

Marsh Life *Feathers and Fur*

One afternoon as I was watching our two little boys chase fiddler crabs at the edge of a large mudflat separating two stands of *Spartina*, I noticed a clapper rail nervously standing at the edge of the marsh. The bird darted gingerly over the soft mud to the other stand of marsh grass. A minute later, another clapper rushed out of the *Spartina* stand and followed the first into the grass on the other side of the mudflat. Then the two rails repeated the performance, running back one after the other to the original *Spartina* marsh. Whether they were a courting couple or rivals, I never knew. They didn't come out again.

Both birds and beasts use the resources of the salt marsh. Some, like clapper rails and rice rats, are secretive and seldom seen. Others, including herons, egrets, raccoons, and deer, are often spotted while scouting out a meal in the salt marsh or along its borders. The marsh offers refuge and food to many feathered and furry vertebrates as well as to the scaly fish and reptiles.

Birds

Bird watching is an enormously popular activity in the United States, and the study of birds is popular among ecologists as well. Birds are fun to spot and watch because they are active during the day; have distinctive, often brightly colored feather patterns; and make interesting and easily recognized songs and calls. Most paleontologists agree that birds are the last living descendants of a branch of dinosaurs, the theropods, or raptors, which include *Tyrannosaurus* and *Velociraptor*, both featured in the movie *Jurassic Park*. Fossils of Cretaceous raptors have clear impressions of feathers around the bones. These predators didn't fly, so it is likely that their downy coats conserved body heat and may have been patterned to attract mates. True birds first appeared earlier, in the Jurassic period. Bird diversity expanded during the Cretaceous and then exploded in the Tertiary, along with the diversity of mammals, after dinosaurs had exited the scene.

At any rate, the southeastern coast, with its great variety of relatively wild habitats for waterfowl and other species, is one of the premier birding spots in the country. A large number of birds are annual residents, and many others migrate through or spend part of the year on or around the sea islands. More than three hundred bird species have been sighted along the Georgia coast. The State of Georgia has highlighted this avian extravaganza by designating the Colonial Coast Birding Trail, which pinpoints numerous birding hot spots, from the north beach of Tybee Island to the wilds of the undeveloped Cumberland Island National Seashore.

Some coastal birds live entirely in the salt marsh. Clapper rails, *Rallus longirostris*, also called marsh hens, are medium-size grayish-brown birds with long bills. Abundant in *Spartina* stands, rails are usually heard rather than seen; their loud, raucous call is a common sound of the marsh. They do occasionally emerge from the shelter of the cordgrass stalks, as I found out that day in the marsh. The preferred foods of clapper rails are squareback and fiddler crabs, but ribbed mussels (despite potential toe loss!) and periwinkle snails are also part of their diet. Rails lay their eggs in the marsh grass, building woven nests in the tops of stalks that rise above all but the highest tides. The eggs can survive an occasional inundation with saltwater. There is a fall hunting season for marsh hens. I was given a baked rail once. It looked like a naked pigeon and tasted gamey. I decided that I prefer my clapper rails running over mudflats between patches of cordgrass.

Two other marsh-dwelling birds are as secretive as rails. The most common bird of tall creekside *Spartina* is the long-billed marsh wren, *Cistothorus palustris*. Like most wrens, they are shy and difficult to spot, but in spring and early summer the melodious songs of the males fill the marsh. Marsh wrens feed on spiders, grasshoppers, and other insects in the cordgrass. During breeding season, the male builds several flimsy nests among the *Spartina* leaves. When the female chooses a mate, she ignores his pitiful efforts and starts a final, well-constructed nest that the male dutifully helps finish. Egg laying must be timed to occur between the highest spring tides so that the young birds are hatched and fledged before the nest is flooded; submersion in saltwater would kill the nestlings.

Sharing the wren's habitat is the seaside sparrow, *Ammodramus maritimus*, a little gray bird with a thin yellow marking above the eye. The two species nicely divide the resources of the marsh. While the wren hunts insects and spiders in the canopy of creekside cordgrass, the sparrow seldom flies, but runs about on the marsh surface after small crabs, worms, and other benthic invertebrates. At first glance, the little bird scurrying through the *Spartina* stalks may be mistaken for a mouse. Herbert Kale, who wrote a monograph on the marsh wren in Georgia, commented: "I never collected a marsh wren with muddy feet, or a seaside sparrow with clean feet." The sparrows choose nesting sites in the short cordgrass stands of the upper marsh. A close relative of the seaside sparrow, the sharp-tail sparrow, *Ammospiza caudacuta*, breeds in northern salt marshes during summer and overwinters in southeastern salt marshes.

In the fall, great flocks of migrating tree swallows, *Tachycineta bicolor*, swirl over the salt marshes in search of insects. One October day as I was taking samples in a short *Spartina* marsh, a large cloud of hundreds of tree swallows flew right over my head. For a long few moments I was completely surrounded by the rapidly beating wings and high-pitched twitters of the little birds. I thought of the Hitchcock movie; then they were gone. Tree swallows also congregate around beach dunes, where they supplement their diet with the fall crop of wax myrtle berries.

The edges of the marsh resound all year with blackbird calls. Large, noisy, iridescent birds with long rounded tails, boat-tailed grackles, *Quiscalus major*, flock together with smaller red-winged blackbirds, *Agelaius phoeniceus*. When I first came to Sapelo as a novice birder, I was puzzled by the smaller brown birds that always seem to hang out with the dark, handsome grackles and red wings. Then I consulted the field guides carefully and discovered these were female blackbirds. Both the grackle and the red-winged blackbird have unforgettable calls. One particular ascending, high-toned grackle sound seemed to me just like that of the handheld communicator used in the classic *Star Trek* TV series.

The most beautifully colored songbird of North America is abundant on the sea islands in summer. The male painted bunting, *Passerina ciris*, with a bright blue head, green back, and fire-engine-red body, looks like it flew out of an artist's palette. In early April, the males arrive from tropical wintering grounds in South America. The oak trees, groundsel bushes, and dune shrubs are suddenly filled with melodious bunting claims to nesting territory. By May, the yellow-green females have shown up to pair off with successful male claim stakers. During one late spring field trip to Sapelo, a teaching assistant, pointing out various bits of natural history of the island to a class of college students, mentioned that they should watch for these beautiful little birds. One student, who had obviously not been paying careful attention to the assistant's comments, started looking excitedly around and exclaimed: "I sure do want to see the painted bunnies!" Often seen along with the painted buntings are electric-blue indigo buntings, *Passerina cyanea*, which nest in the bushy borders of salt marshes and in dune vegetation.

During fall and winter, myrtle thickets along the marsh edge and in the back dunes of the beach are filled with enormous flocks of small dark-gray birds with bright yellow wing and rump patches. These are myrtle warblers, also known as yellow-rumped warblers, *Setophaga coronata*, which find that the mild temperatures and abundant myrtle berries of the Georgia coast make a perfect place to spend the cold part of the year. These cheerful little birds are fairly tame; by standing near a myrtle bush and softly going, "Scheep, scheep, scheep," somewhat like a baby bird trying to get its parents' attention, you can soon see myrtle warblers hopping in the branches all around you.

The king of the salt marsh birds is the great blue heron, *Ardea herodias*, a magnificent blue-gray creature sporting a black plume on its white head. The great blue heron stands over three feet high and in flight has a six-foot wingspan. Its chest is adorned with a billowy tuft of feathers. When hunting in the marsh during low tide, the tall heron regally stalks up the mudflat of a shallow tidal creek, bright yellow eyes searching for movement in the water. When the bird suddenly freezes, it has spotted a potential meal. An instant later, the long yellow bill stabs into the creek after a fish. One might think that herons impale their prey with their sharp beaks. Actually, the bird grabs the fish between its two mandibles and then, with an experienced toss of the head, flips up the catch and swallows it headfirst to avoid the fins sticking in its throat. Great blue herons have a varied diet: fish and crabs in the marsh, and frogs, snakes, and small rodents in the uplands. I have seen a photo of a great blue swallowing a rather large rat. Although the great blue is an impressive sight, like all herons it has an unlovely voice. When startled, the big bird flies away, complaining with loud, hoarse "craawks."

Other herons share the great blue's kingdom in the marsh. The little blue heron, *Egretta caerulea*, is only half the size of its big cousin. Immature little blues are pure white, while the adult plumage is completely dark blue with a reddish head and neck. The black-tipped light-colored beak is distinctive. The tricolored heron, *Egretta tricolor*, is similar in size to the little blue heron, but it has a dark head and white underparts. These smaller, wading herons are more agile feeders than the ponderous great blue, but their efforts pale beside the show given by the reddish egret, *Egretta rufescens*. These shaggy red-brown herons attract small fish by holding their wings outstretched over calm water. The fish congregate in the shadow of the wings, thinking that they have found safe shelter, until the egret starts feasting on its catch. In addition, reddish egrets jump and flap their wings over the water in order to herd schools of fish toward shallow spots in tidal creeks, where the hunting is easy. These herons, which are more common along the Gulf of Mexico coast, are not often spotted in Georgia.

The smallest heron that hunts in tidal creeks is the green heron, *Butorides virescens*. Despite its common name, this bird is more blue than green. It has a reddish-brown neck and bright yellow or orange legs. The little green heron is as much at home in small freshwater ponds as in the marsh. In either setting, it always seems to be perched, still as a statue, on an oyster reef or a branch at the edge of the water. I don't remember ever seeing a little green catch anything, although judging from their abundance, they must be pretty successful hunters.

While fishing on a dock at night, more than once I have been startled by a sudden loud croak and the flapping wings of a heavy bird launching from a nearby piling. Invariably, it turned out to be a black-crowned night heron, *Nycticorax nycticorax*. Unlike other herons, this one has a short thick neck and short legs, and is active at night rather than during the day. The night heron's plumage is attractive: dark above, white below, and it has red eyes and yellow legs. While on Sapelo, I often found myself staring for some time at a funny lump on a tree branch overhanging the pond by the lab and then realizing it was a night heron roosting during the daytime.

The loveliest coastal birds are the graceful egrets. These herons are named for their long plumes of feathers (*aigrettes* in French). Egrets were hunted nearly to extinction in the late 1890s and early 1900s, their feathers used to adorn ladies' hats. Luckily, the nascent Audubon Society and President Theodore Roosevelt acted to end the killing of egrets and to create bird sanctuaries around the Atlantic and Gulf Coasts. The great egret, *Ardea alba*, is the largest white bird seen in the salt marsh. About the size of the great blue heron, the great egret has a yellow bill and legs. The smaller snowy egret, *Egretta thula*, can be recognized by its black bill and black legs with yellow feet.

When our family lived on Sapelo, the pine trees growing around the Marine Institute were a favorite resting place of the snowies. At the end of the day, flocks of these egrets would fly in for their evening roost. One summer, the egrets seemed to choose a certain low-growing, bushy pine in the middle of the pond as the favored roosting site. Our young sons clamored to go see the "bird tree" after supper. Hundreds of snowy egrets would be perched all over the poor little tree; its branches were festooned top to bottom with their droppings. Social status in the egret community must have determined roosting position. The birds in the top branches always appeared haughtily at

Snowy egret stalking killifish next to an oyster bed in a marsh creek mudflat at flood tide.

ease, while quite a bit of jostling for space went on at the lower elevations. Every now and then an ousted egret would flap off with loud protest to roost in some less fashionable tree.

Cattle egrets, *Bubulcus ibis*, are just a bit smaller than snowy egrets and have a yellow bill. In breeding season, adults sport orange patches on the head, breast, and shoulders. Cattle egrets are native to Africa, where they follow after gazelles, zebras, and wildebeest, feeding on insects the grazers stir up from the grass. Somehow, early in the twentieth century, cattle egrets managed to cross the Atlantic Ocean to South America, likely riding equatorial trade winds. They readily adapted to life in pastures with domestic cattle, and gradually worked their way northward to the United States. In a brief note to the ornithological journal *Oriole*, John Teal recorded the precise date that cattle egrets were first spotted on the Georgia coast: June 6, 1956. On that day he saw two adults and a juvenile stalking behind Reynolds' cows in an open field on Sapelo Island. Teal and other ecologists had been on the lookout for these avian immigrants, which by the 1950s had become common in Florida, but none were reported in Georgia until that date in 1956. In the summer of 1957, Teal noted several more cattle egrets and observed a nesting pair close to a heron rookery in a pond at the north end of the island. Now cattle egrets are abundant all over the Southeast. Since there are no longer herds of cows on the island, the egrets have shifted to hunting crabs and grasshoppers in the salt marsh meadows.

The other white birds commonly spotted along the coast of Georgia are white ibis, *Eudocimus albus*. Adults are all-white with a red face and a long curving bill; immature white ibis are brown above and white below. While herons tend to be solitary feeders, ibis are gregarious, hunting small invertebrates in coastal marshes and freshwater ponds in groups of a dozen or more. Pine and cypress trees around freshwater ponds on Georgia sea islands are popular sites for both heron and ibis nesting colonies, or rookeries. John Teal noted about a thousand white ibis nests in the summer of 1958 in ponds on the north end of Sapelo.

Glossy ibis, *Plegadis falcinellus*, uniformly reddish brown in color, are less common in the Southeast. I only rarely saw glossy ibis while on Sapelo, but they are increasing in abundance. In flight, ibis fly with necks fully extended, in contrast to herons and egrets, which fly with their necks held in a graceful S shape.

One of the most breathtaking birds in this region is the wood stork, *Mycteria americana*, also called the wood ibis by early American ornithologists because it has an ibis-like profile in flight. This is the only stork native to North America; the five-foot wingspan, white body with black wing margins and tail, and dark head are unmistakable. Wood storks often soar high above the salt marsh or take a daytime break from feeding by roosting in live oak trees growing on marsh hammocks. Wood storks nest in huge rookeries in cypress swamps along the lower Altamaha and Savannah Rivers. There the birds build their flimsy stick nests and rear one or two young, safe from human disturbance.

The belted kingfisher, *Megaceryle alcyon*, is a familiar sight on power lines and dead tree branches overhanging salt marsh creeks, patiently waiting to spot a small fish in the water below. Then the kingfisher abruptly does a "power dive" into the creek and emerges, more often than not, with a struggling mummichog or young mullet grasped in its bill.

Another predatory bird that is engaging to watch is the harrier hawk, *Circus cyaneus*. This medium-size brown raptor with white underparts, a diagnostic white rump, and a long tail flies in endless low circles over the *Spartina* marshes, prowling for small birds and rodents.

Ospreys, *Pandion haliaetus*, can be spotted in Georgia estuaries during their spring and summer nesting seasons. While most hawks are dark, rather heavy-bodied birds, fish-hunting ospreys are lighter in color and have a slimmer profile. There was a severe decline in osprey numbers during the 1960s due to use of DDT and other pesticides, but since then their populations have rebounded.

When I lived on Sapelo, the shrill "kree-kree" osprey scream was often heard over the salt marshes in summer. A pair of ospreys nested every year on one of the tall power-line poles on the bank of the Duplin River. One summer in the late 1970s, I passed their nest often as a colleague and I motored up the tidal river to take routine samples. In early spring I noted that the adults had returned to the pile of sticks on top of the pole, all that winter storms had left of the previous year's nest. Once the nest was satisfactorily repaired, the adults began taking turns sitting in it. Usually, all I could see of the osprey in the nest was its beak and a watchful eye as our little boat sputtered by. Apparently, at least two eggs were being incubated, since a few weeks later a couple of scrawny chick heads would appear at the edge of the nest as we passed. One or the other of the parents would hover over the chicks with warning screeches until we were gone. More often than not, we saw the other adult returning with a mullet in its talons to supply the ever-hungry nestlings. By midsummer, the young ospreys were fully fledged and sitting on the edge of the nest or on the power line. Soon they would be fishing for themselves. A few years later, I noticed a second osprey nest, somewhat inexpertly built, on top of a power-line pole several places down from the original nest. I presumed that one of the osprey fledglings had returned with its own mate to raise a family on good Duplin River mullet.

The only predatory bird larger than the osprey seen on the Georgia coast is the American bald eagle, *Haliaeetus leucocephalus*, which also suffered a severe population decline in the middle of the last century because of pesticides. Old-timers on Sapelo remembered when bald eagles nested among the herons and ibis in the large freshwater pond at the north end. During my time on Sapelo, the Georgia Department of Natural Resources started feeding imported adolescent eagle chicks in a specially constructed nest tower in this same pond. The hope was that as adults, the eagles would return to a familiar place to rear their own young. This program was later expanded to other sites along the Georgia coast, and by the early 2000s there were dozens of eagle nests. Bald eagles are now often sighted soaring over the estuaries.

The vast open marshes and long stretches of ocean beach, where dead animals are prone to be carried by the tides, are ideal for scavenger birds. Soaring buzzards and flocks of crows are omnipresent. Black vultures, *Coragyps atratus*, are often seen along the coast. These large dark birds, with their characteristic, teetering V-shaped flight profile, search the marshes and the wrack line of the beach for carrion; they rest with hunched wings on branches overhanging the tidal creeks. Glossy black fish crows, *Corvus ossifragus*, scavenge around marshes and beaches. Both vultures and fish crows, despite their reputations, rely on more than dead creatures for food. Vultures can bring down live game, and fish crows fly inland to feed, along with other crow species, on farmers' crops.

During my time on Sapelo there were three musts for a resident: spotting some of the wild cows left behind from the R. J. Reynolds cattle herds of the 1950s, which roamed in small bands on the north end of the island; finding a note in a bottle washed up on the beach; and catching sight of the elusive colony of chachalacas on the south end. It wasn't long before I saw the cattle during a truck ride up north; my first sighting was of two long-horned bulls that looked as big as rhinoceroses. I also found a bottled note on the beach (I don't remember what the message was). But I was frustrated for quite a while on the third count. Chachalacas, *Ortalis vetula*, are arboreal pheasants common in Central America. In the 1920s, Howard Coffin released some of these pheasants on Sapelo Island to give his pals something more challenging to shoot than wild turkeys. In *Portrait of an Island*, their narrative of the natural history of Sapelo, the Teals mentioned a healthy breeding population of chachalacas during the late 1950s. (The *Breeding Bird Atlas of Georgia* confirms that in 2009 the birds were still happily inhabiting the live oak groves on Sapelo island.)

The olive-brown long-tailed pheasants are both gregarious and garrulous. I heard them often in the woods around the Marine Institute. Their loud, gabbling, eponymous "cha-cha-ah-cah" was hard to miss. I was able to check chachalacas off my list of musts one evening as I was driving back from Nannygoat Beach. Startled by a brown flapping body hurtling over the hood of the truck, I stopped and looked back in time to see several more long-tailed birds launch themselves from a tree on one side of the road and sail over to a teetering perch in a chinaberry bush on the other side. After I had a brief glimpse of the flock in the fading light, they disappeared farther back into the bushes. A moment later, the teasing chorus began: "cha-cha-ah-cah, cha-cha-ah-cah, cha-cha-ah-cah."

Mammals

Although several mammals regularly use the salt marshes or beach dunes as places to feed or even to raise their young, they are mostly nocturnal and wary, and not often spotted. Marsh rice rats, *Oryzomys palustris*, prowl *Spartina* marshes, especially along creek-bank levees. These gray-brown rodents with cream-colored undersides were once a major pest in lowland rice plantations in the Southeast, hence their common name. Marsh rice rats are not picky about their diet, though fiddler and squareback crabs are mainstays on their menu. Cordgrass seeds and juicy stems are carbohydrate-rich additions in the fall. Rats are seldom noticed during the day, but their tracks can be seen in the soft creek-bank mud from the previous night's foraging. Abandoned marsh wren nests are favored by rice rats for their daytime naps. In a flagrant abuse of this hospitality, marsh wren eggs are a rice rat delicacy. In fact, predation by rats on wren nests may be a factor limiting the population size of these marsh birds. Rice rats build nests of grass in the upper marsh to raise their litters. With a typical life span of just six to twelve months, female rats are amazingly fecund; during their short lives, they manage to produce up to six litters, with an average of five young in each.

Another coastal rodent is the cotton rat, *Sigmodon hispidus*, found in stands of *Spartina patens* and in wax myrtle thickets on marsh borders. The eastern wood rat, *Neotoma floridana*, and old field mice, species of *Peromyscus polionotus*, are locally common in the back dunes of the beach, where they feed on the seeds of sea oats and other dune plants. These small rats and mice are prey for hawks and rattlesnakes that hunt the marsh and dunes.

The marsh rabbit, *Sylvilagus palustris*, lives in the dunes and marsh edges on sea islands and in the brackish marshes of river deltas. Marsh rabbits forage for succulent plant stems well out onto the tidal marsh, becoming active about an hour before sunset. The marsh rabbit is adapted to saline habitats; in similar freshwater environments on the coastal plain, a related but separate species, *Sylvilagus aquaticus*, occurs.

When our family lived on Sapelo, our next-door neighbor's golden retriever, Claw (short for Crabclaw), occasionally seemed to go nuts. She would stand out in the backyard with her ears cocked to the ground and then suddenly pounce on a patch of grass and start digging away like crazy. When we saw her bring up a little brown body from the hole, we knew she had heard the soft scraping of a mole burrowing under the lawn. The eastern mole, *Scalopus aquaticus*, is a very abundant mammal of sea islands but, paradoxically, is very rarely seen. One finds mole burrows anywhere the soil is suitable for these subsurface tunnelers, including beach dunes and even sandy tidal flats. Moles feast mainly on earthworms and insect larvae they come across underground, but supplement their diet with starchy plant roots.

The best-known mammal frequenting the salt marsh is the raccoon. Often seen during the day along the coast, raccoons travel far into the marsh at low tide to feed on fiddler crabs. The eggs of marsh birds and terrapins are a special raccoon treat. The Teals described how raccoons survive when caught out on the marsh at flood tide: these clever animals quickly construct a dry nest by folding down cordgrass stalks and then rest until the water leaves the marsh. Raccoon droppings, rich in the white shelly fragments of crabs, often litter docks, as if raccoons prefer a scenic view while answering the call of nature.

Raccoons also seemed to fancy our little sailboat, the *Sea Squirt*, which was moored to a floating dock up a tidal creek. Once, after we had spent a day washing the boat in preparation for a weekend sail, the raccoons must have decided to celebrate our clean boat by throwing a party on board that night. When we arrived bright and early the next morning to depart, muddy raccoon paw prints were plastered from stem to stern, even way up into the rigging. Fiddler crab claws and carapaces were scattered all over the cockpit and cabin roof. After another hour of scrubbing while muttering at the inconsiderate buggers, we got underway. Then came the final raccoon salute: when we raised the mainsail, several large droppings fell out of it and splattered all over the newly cleaned deck.

Two other carnivorous mammals that hunt in salt marshes are mink and river otters. These slim dark members of the weasel family frequent creek banks and the low salt marsh, searching for crabs, clams, and fish. For sleeping and raising their families, they prefer dry land at the edge of the marsh. Before they were protected by law, American mink, *Neovison vison*, were heavily trapped for their soft pelts; today they are rare in Georgia marshes. I saw a mink only two or three times during my years on Sapelo. Early in the morning or late in the afternoon while waiting for the mainland ferry, I glimpsed a slender dark animal poking around in the oyster reefs on the mudflats around the dock.

One day when I was gazing out a back window of the Marine Institute around midday, I noticed what seemed to be a long-bodied, low-slung black dog sauntering down the main road to the building. It didn't match any of the island pets I knew about; then I realized it was an otter. River otters, *Lontra canadensis*, are fairly common in undisturbed areas of the estuary. They may be spotted on marsh banks or swimming in tidal creeks. The antics of otters at play are captivating. A friend told me of watching three otters wrestling and splashing happily in a marsh creek.

The largest land mammal on the sea islands is the white-tailed deer, *Odocoileus virginianus*. Island deer are smaller than those on the mainland. There are even separate subspecies of whitetails on particular barrier islands. The deer graze on marsh grasses; at the edge of the marsh one sometimes comes across a nest of matted-down plants within the bordering bushes, indicating a place where a deer rested. Island deer also like the beach dunes, where they find both food and shelter.

During the Teals' stay on Sapelo in the 1950s, island deer mainly kept to the north end of the island and were not often seen around the Marine Institute. In my time, however, the Georgia Department of Natural Resources held fall hunts on the north end of Sapelo. The deer seemed to have discovered that the southern part of the island, outside the hunt boundaries, was safer. The deer were still wary, but I often saw small herds grazing on lawns and on acorns under the live oaks, even during the day. Island gardeners surrounded vegetable patches with electrified fencing to keep the deer out. One summer, my family noticed a buck hanging around our garden in the late evening. He tore down a section of our fence while trying to get in one night. But the shock did not deter him. One morning a few days later, all our pole beans and the tops of the tomato bushes were gone. The fence was intact, so the buck must have figured out he could jump over the wire. Deer can easily spring over fencing as high as six feet. I didn't begrudge him, though. One misty morning l looked out the window of our living room and saw a doe and her spotted fawn grazing on the grass just a few feet from the house. Having deer feel comfortable being that close to us was worth some beans and tomatoes.

One land mammal is a very recent addition to Georgia sea islands: the armadillo. The nine-banded armadillo, *Dasypus novemcinctus*, has been expanding its range from Central America and Mexico since about 1850. It had crossed the Rio Grande and populated the hills of South Texas by the 1880s (it has been designated the official small mammal of that state). Within a few decades, armadillos were burrowing into fields and forests from Oklahoma to Mississippi. Armadillos were not well established in Georgia by the time our family left in 1990. When we returned to Sapelo for a visit in 1999, we had a shock. There were serious armadillo excavations in the scrub all over the island. The armored invaders had arrived in force.

For these creatures, which are most closely related to sloths and anteaters, water is not much of an obstacle to their migrations. Armadillos can inflate their guts with air to make them buoyant enough to float over a river. To get across shallow tidal creeks, they exhale, sink to the bottom with their bony armor, and scrabble over the creek bed underwater. Armadillos have few natural predators in their new habitats, they are omnivorous, and they can breed like crazy. Females have their first litter when just a year old. Curiously, each litter starts out as a single fertilized egg, which, on implantation in the mother's uterus, divides and develops into four identical embryos. So, each armadillo brood is either all-male or all-female quadruplets.

Now that we have surveyed the inhabitants of the salt marsh and tidal creeks, let's follow the flood tide back to the watery realm of the estuarine rivers and sounds.

9
What Lies Beneath *Zooplankton*

Standing on a dock, you watch the sea current flow by, dark and mysterious. Sydney Lanier must have felt the same, when he lamented in his famous poem:

> *And I would I could know what swimmeth below when the tide comes in*
> *On the length and the breadth of the marvellous marshes of Glynn.*

But thanks to the abiding curiosity of scientists who have scooped and netted and trawled in the sounds and tidal rivers along the southeastern coast, we do know what lies under the wind-rippled skin of the estuary. We know a great deal, in fact, about the animals that "swimmeth below."

Let's start with zooplankton (Greek for "animal wanderers"), species of invertebrate that are weak swimmers, and are carried by the waves and tides throughout the estuary. Georgia estuarine waters harbor great throngs of zooplankton that drift about and feed on the even smaller algae and predatory protists. Zooplankton are prime food for small fish and for the larvae of fish and shellfish, some of which are themselves part of the plankton. A few types of zooplankton—for instance, jellyfish—are conspicuous, but most are not easily seen unless they are concentrated with a fine mesh net (see appendix 3 for instructions on making a plankton net out of common household supplies). Representatives of almost every major group of invertebrates can be found in the zooplankton, in either adult or larval form.

Several distinct types of zooplankton are mainly translucent and resemble lumps of clear gelatin when removed from the water. Not surprisingly, marine biologists refer to all these creatures as "gelatinous zooplankton." The most common of these are jellyfish, in the phylum Cnidaria, which are among the most primitive animals on earth. Two layers of skin cells sandwich a clear jelly-like padding that gives shape to their bodies. Their simple nervous system lets them detect and capture prey with their tentacles, which are loaded with nematocysts, specialized stinging cells that discharge long threads full of toxic venom when the tentacles encounter a potential meal. Modern cnidarians include benthic anemones and the polyps of corals.

Two other cnidarian groups pulse about in the plankton: the bell-like jellyfish and the box jellies. Jellies with circular bells are the familiar ones featured in riveting displays in marine aquariums. Watching as the alternately contracting and expanding bells trailing long tentacles circulate up and down in the tanks is a Zen-like experience, as meditative as staring at rising and falling blobs in lava lamps. On the other hand, box jellies, with bodies that are cubic rather than round, are rarely kept in aquaria, for good reason. Box jellies are among the most venomous animals on earth. These transparent jellies, although generally smaller than their bell-shaped kin, are strong swimmers, can travel at rates of up to eighteen feet a minute, and have complex eyes, complete with lenses and retinas. Their visual acuity and speed make them active hunters of small fish. Their potent toxins stun their struggling victims into quick submission. The famous box jellies that swarm off the eastern and northern coasts of Australia cause agonizing stings that have even killed badly stung swimmers by causing cardiac arrest in a matter of minutes.

The simple body plan of gelatinous zooplankton suggests that these animals evolved early in earth's history. Indeed, the fossil record shows that jellyfish were among the first animals to appear in marine plankton. Since they don't have bones or shells, jellyfish are not easily detected in ancient rocks. Fortunately, animals that died and fell into soft marine muds often left impressions of their bodies in mudstone formations. Paleontologists had long been teased by fossil forms resembling the bells of jellyfish in shale rocks that formed from marine muds 500 million to 600 million years ago, when multicellular animals first appeared in the sea. But scientists couldn't be sure that these were the ancestors of modern cnidarians. Then, in 2007,

Paulyn Cartwright, a professor at the University of Kansas, reported that she and her colleagues had discovered beautiful fossil imprints resembling modern jellyfish, complete with bells and tentacles, in mudstone deposits in Utah. The rocks dated to the middle Cambrian, 505 million years ago. Surprisingly, impressions of both bell-shaped and box-shaped forms were discovered, suggesting that both these groups of planktonic jellies had evolved by that time. Oceanographers who study the geological history of the sea call this primordial period the "jellyfish ocean," since these creatures were the top predators in the plankton before the appearance of fish. If the Cambrian box jellies had venom as lethal as that of modern species, they certainly could have subdued the trilobites and other creatures that swam in those ancient seas.

There is now serious worry that the ocean is returning to a jellyfish-dominated state. (The unnerving book *Stung! On Jellyfish Blooms and the Future of the Ocean*, by Lisa-ann Gershwin, details this concern). Outbreaks of jellyfish blooms have been alarming fishermen and beachgoers in recent years. One of the most amazing developments has been the proliferation of the giant jellyfish, *Nemopilema nomurai*, in the China Sea. The bell of this enormous jelly grows up to seven feet across, and a large specimen can weigh several hundred pounds. First identified in 1921 by a Japanese fisheries researcher, Kanichi Nomura, until recently this species was a marine curiosity. Lately, however, vast swarms of Nomura's jellyfish have been migrating from their traditional habitat in the warm China Sea up the coast of Japan during summer. Considering the weight of just one of these monsters, it is no wonder that the nets of fishing boats are clogged and broken by these jellies. Smaller jellies, such as the mauve stinger, *Pelagia noctiluca*, have been carpeting the Mediterranean Sea and keeping tourists away from beach resorts.

Exceptional jellyfish blooms are thought to result from a combination of sea warming (due to climate change) and the overharvesting of fish that are jellyfish predators and competitors for plankton food. Both warmer waters and fewer fish would boost jellyfish numbers, and these simple, brainless bags of goo are eating machines that can grow like crazy when they get a chance. It might be, though, that jellyfish go through natural boom-and-bust cycles, so the scientific jury is still deliberating whether these creatures are going to retake the sea. It is just as well, though, that the enormous, net-busting Nomura's jellyfish doesn't live off the Georgia coast.

Although no deadly or giant jellies occur in southeastern estuaries, many species of these primitive animals, some stinging and some rather large, are found here. Two species of stinging jellyfish are especially common. The lion's mane jelly, *Cyanea capillata*, has a pinkish-red umbrella, or bell, with frilly tentacles in eight separate clusters. Along the warm southeastern coast, it rarely grows to more than five inches across, but in colder northern waters the bell of the lion's mane can reach several feet in diameter. The largest documented lion's mane was found on a beach in Massachusetts in 1870; the bell was seven and a half feet across, and the tentacles were 120 feet long. So the lion's mane is actually on record as the largest jellyfish in the ocean, even bigger across than the monstrous jellies found off Japan. Since the lion's mane prefers colder waters, it is most abundant along the Georgia coast during winter.

The sea nettle, *Chrysaora quinquecirrha*, has a thick flat bell, up to six inches across and is white to reddish brown in color, with about twenty tentacles hanging from the margins. Its tentacles are loaded with nematocysts. The sea nettle is a voracious predator of zooplankton and larval fish. This species, unlike the lion's mane, prefers warm water. Also unlike other jellies, the sea nettle tolerates brackish water, which allows it to thrive in estuaries and up river mouths. Each summer, sea nettles swarm in Chesapeake Bay, stinging swimmers and feasting on baby fish.

Sea nettles are also a problem along the Georgia shore. Researchers diving at Gray's Reef, a hard-bottom habitat several miles offshore from Sapelo Island, have reported such dense concentrations of nematocyst-laden sea nettles and other jellies that they could not avoid being constantly stung. They are also a danger for swimmers at the beach. Sea nettles' tentacles break off in the surf, so that even when jellyfish bells are nowhere in sight, a painful sting can be delivered by an unseen free-floating tentacle. Jellyfish washed up on the sand can zap a toe or finger. The recommended treatment for a jellyfish sting is to wash the affected area with saltwater, not freshwater, since freshwater will cause the nematocysts to release more toxins.

One of the largest jellyfish along the southeastern coast is the cannonball, *Stomolophus meleagris*. Locals call them "jellyballs." Unlike its flimsier cousins, the cannonball has a very firm, pinkish-white bell, up to a foot in diameter, with a mass of dark reddish-brown tentacles protruding from the bottom. When discharged, cannonball nematocysts are shorter than those of the lion's mane, so they can't penetrate human skin very well. I nonetheless managed to get stung once by a cannonball. Leaning far over the front end of a Boston whaler to scoop one out of the water, I toppled over and found myself face-to-bell with it. I felt a slight stinging around my eyes, but that was all. Young spider crabs often hitch a ride on these big jellies, finding shelter and free meals until they are large enough to compete with adult crabs on the sea bottom. The juvenile crabs stay with their host even when the jellyfish is washed up onto the beach. When I picked up one cannonball from the surf, no less than five inch-long spider crabs were crawling around inside the bell.

On research cruises off the Georgia coast, I have seen row after row of thousands of cannonballs pulsing along the surface. To fishermen, the cannonball jellies are a nuisance. During the summer, the abundance of cannonballs is so high that shrimpers' nets become clogged with them. Their trawls are so often fouled with the big jellyballs that they have been forced to add mechanical "jellyball shooters" to their nets to shunt them out of the way. But one clever fisherman knew that dried jellyfish was a delicacy in China and Japan. He decided to set up a cannonball processing plant in Darien, a fishing port on the Altamaha River. By the early 1990s, shrimp fisheries along the Georgia coast had become less profitable because of a flood of cheap imported farmed shrimp. Catching and processing jellyballs for export had the potential to supply a niche market, possibly bringing in more money than shrimp trawling. The jellyfish-processing plant in Darien has been operating ever since. The cannonball jellies are easy to catch by slow trawling. The maximum quantity of jellyballs that can be processed each day by the Darien facility is sixty thousand pounds. One fishing boat can get that much in a night. Once delivered to the factory, the bells of the jellies are dried, packaged, and shipped out to Asian countries. So far, there is not much demand in the United States for dried jellyfish. Even the Darien jellyball fishermen who have tried the stuff vow that once is enough. But jellyballs are currently the third-largest fishery product, by weight, in Georgia.

The most numerous gelatinous zooplankton are tiny compared to the cannonballs and sea nettles, and so usually go unnoticed. When the catch of a plankton tow is held up to the light, there will likely be transparent bells less than an inch or two across, pulsing about. These hydromedusae jellies have a complex life history. Typically, there is a sessile (attached) hydroid stage that grows on the bottom or on a hard surface as a fleshy polyp. The hydroid produces medusae, miniature jellyfish that live, grow, and reproduce in the plankton, creating yet another hydroid stage, which settles to the bottom. Some species apparently have learned to do without the hydroid, retaining only a planktonic medusa stage. The small translucent jellies feed on copepods and other zooplankton.

A distinctive jellyfish relative is the Portuguese man-of-war, *Physalia physalis*. Unlike true jellyfish, the man-of-war is actually a colony of individual animals. The bluish-purple translucent bladder, up to a foot long, rises above the waves and acts as a sail to move the colony about in the sea. The stinging tentacles that trail below can reach fifty feet in length. This open-ocean jelly is occasionally carried up onto beaches after storms; caution is advised if one is found, since the very painful nematocysts remain active for quite a while.

Comb jellies are curious creatures in their own separate phylum, Ctenophora. Although they are gelatinous zooplankton, they don't have nematocysts and can't sting. Ctenophores are generally transparent, cylindrical in shape, and swim with eight rows of hairlike cilia that beat in waves. The cilia lined up along the body resemble the teeth of a comb, hence the common name (likewise, "ctenophore" is Greek for "comb-bearing."). Like sea nettles and other jellyfish, comb jellies are voracious predators of copepods and other crustacean zooplankton, which they actively hunt and swallow whole. The most common ctenophore in southeastern coastal waters is Leidy's comb jelly, *Mnemiopsis leidyi*. Its clear oval body is one to four inches long. *Mnemiopsis* is bioluminescent, making its own light in the same way that fireflies do. During summer nights, this ctenophore is the source of brilliant white flashes in the water. Taking a cruise across an open sound in the estuary on a dark summer night is a treat. The twinkling of comb jellies within the soft blue fire of dinoflagellate bioluminescence in the boat's wake echoes the sparkling stars above.

By far the most abundant zooplankton in the sea are tiny crustaceans, most of which are copepods. When captured with a plankton net, copepods are just visible as whitish specks about the size of a sesame seed or a grain of rice, darting madly about. With two long antennae that they use to sense prey or predators, and multiple swimming legs, copepods are perfectly adapted for a life in the plankton, feeding on algae and protists. Lipid-rich diatoms are a favorite food. Copepods store extra fat from their algal prey in their bodies as sacs of waxy oil. Their lipid stores make copepods a nutritious, high-energy food for fish ranging in size from tiny larvae to filter-feeding basking sharks. Although they can't make headway against tidal currents, copepods can swim well enough to migrate up and down in the water.

Common copepods in Georgia coastal waters are the calanoid species *Acartia tonsa* and *Parvocalanus crassirostris*. Copepod populations are greatest in summer, when diatoms and other microbial plankton are most productive. During the warm season, copepods can go through their life cycle of egg, several larval stages, and breeding adult in a month or less. Often half or more of the copepod population is composed of the larval nauplii and copepodid stages, which are avidly hunted by small fish.

Many visual predators, such as fish that hunt prey by sight, enjoy a meal of copepods. To avoid these big-eyed hunters, the little zooplankton swim down before dawn to hide in dark bottom waters during the day. After sunset, the copepods migrate up into the surface under cover of darkness to feed. The light of a full moon will send them scurrying back down to the depths. Copepods can sense light, but have no true vision. They seek prey and mates by sensing water movements and chemical odors in their local environment. In many species, female copepods emit an alluring scent that advertises their whereabouts to males.

The crustacean zooplankton also include free-swimming larvae of crabs and shrimp, which often follow the copepods on their daily roller-coaster migrations. So the best time to collect these minuscule crustaceans with a plankton net tow is on a moonless night. Then the wonderfully varied forms of the tiny creatures can be thoroughly appreciated using a hand lens or low-powered microscope.

Along with jellyfish and crustaceans, many other marine invertebrates have planktonic larvae. These larvae assume strange and beautiful forms and usually look nothing like their adult parents. Some of these are big enough to see with the naked eye, but are best observed with the aid of magnification. Common invertebrate larvae include fuzzy polychaete worm larvae (trochophores), butterfly-like mollusk larvae (veligers), and spiny echinoderm larvae (plutei).

Arrow worms, unique members of the marine plankton, are in their own phylum, Chaetognatha, which translates as "bristle jaws." Most coastal chaetognaths are in the genus *Sagitta*, from the Latin word for "arrow." The silvery, slender chaetognath body, less than an inch long, is equipped with a tail and side fins for rapid darting about. Arrow worms aggressively attack copepods, larval fish, and other zooplankton; they use a mouth full of hooked spines to seize their prey, killing them with neurotoxins.

10
Attachment to Place   *Settlers*

As they mature, the free-swimming larvae of some marine invertebrates seek out hard surfaces for a secure place to spend the rest of their lives. They live attached to one spot, feeding on plankton in the water flowing by. Because boat bottoms are favorable real estate for such creatures, the animals that settle and grow there "foul" the vessel, and in aggregate are termed "fouling communities." Our family got to know this marine life well, since we periodically had to haul out our little sailboat and spend an afternoon scraping barnacles and sponges off the bottom. Besides boats, along the Georgia coast there are few sites suitable for fouling animals. Hard surfaces such as pilings, docks, shells, and occasional rocky outcrops are quickly claimed by settling larvae. On the side of a floating dock one can find a treasure trove of marine life: crusty barnacles, bright red and yellow sponges, feathery hydroids and bryozoans, and plump white tunicates. Multitudes of tiny amphipods, worms, and crabs roam over the crowded mass. Other sessile animals find suitable attachment territory in the coastal ocean, on hard sand bottoms, and on the occasional mass of hard rock jutting up from the sea bottom. One such stony outcrop is Gray's Reef, the national marine sanctuary directly east of Sapelo Island.

Sponges are primitive animals that in reality consist of colonies of single flagellated cells that create water currents to bring planktonic microbes close enough to capture and engulf. Sponge cells are related to the minuscule collar flagellates, or choanoflagellates, that live in the plankton. In hard sponges, the flagellated cells of the colony produce tiny spikes of either silica (glass sponges) or calcium carbonate, which form the porous structure of the sponge. Most local sponge colonies are soft, their bodies formed by a tough organic matrix. On docks, one is sure to find a flaming, red-orange mass, usually with "fingers" branching out; this is the red beard sponge, *Clathria prolifera*. Another common sponge, *Halichondria bowerbanki*, ranges in color from yellow to tan and grows as a crust on many kinds of hard surfaces in the estuary. Benthic trawls often bring up a dense gray mass that has a distinctive, garlicky odor when broken open. This smelly object is *Lissodendoryx isodictyalis*, the garlic sponge.

Less obvious are the yellow sponges in the genus *Cliona*, which riddle the shells of oysters and other bivalve mollusks. These shell-boring sponges slow the growth of living oysters, which have to expend extra energy to continuously repair their shells. The shells of dead oysters are completely disintegrated in short order by *Cliona* sponges, the demolition experts of oyster reefs. The calcium carbonate minerals that were in the oyster shells are released into the surrounding seawater and are then available for living oysters to use for the growth of new shells.

Many hydroids are the sessile polyp stage of hydromedusae jellyfish in the zooplankton. The individual animals, called zooids, resemble tiny white flowers growing on a central stalk, which produce free-swimming medusae from buds on the stalk. These colonies are so small that they are easily overlooked. Other species that occur in fouling communities grow only as colonies of polyps, with no planktonic stage. Some of these, such as the bushy mud-colored *Bougainvillia carolinensis* and the long-stalked *Ectopleura crocea*, can appear as large, dense colonies up to a foot tall. The tentacles of each polyp in the assemblage expand into the tidal currents, seeking out copepods and other zooplankton to grab for a meal. Snail fur, a bristly white growth covering hermit crab shells, is another hydroid, *Hydractinia echinata*. *Hydractinia* has special stinging cells, like the nematocysts of jellyfish tentacles, which protect the hermit crab from predators in return for the free ride on the crab's shell.

Corals and anemones, related to jellyfish, are soft, tubular polyps with a crown of tentacles used to grab plankton from the water. Corals are colonies of tiny polyps in a skeleton made of calcium carbonate (hard corals) or in a fleshy matrix embedded with calcareous spicules (soft corals). The soft corals are also known as octocorals because each little polyp has eight tentacles. Two species of soft coral grow on sandy bottoms close to shore, and can be found washed up on the beach. One of these is the sea whip, *Leptogorgia virgulata*, which is made up of many slender, whiplike branches that are bright yellow, orange, red, or purple. Sea whips form dense colonies on shelly sand bottoms and rocky outcrops. They are host to a community of little shrimps and snails that live on and around the colorful branches. These communal creatures often assume the hue of the particular sea whip that they inhabit. After a sea whip dies, the covering of coral polyps erodes away, leaving a flexible black inner core, or whip. The dead whip is a favored site for the growth of bryozoans, and sometimes is completely covered with lacy white crusts or rubbery dead man's fingers.

The other common soft coral found offshore is the sea pansy, *Renilla reniformis*. The sea pansy is a round purple leathery pad one to two inches in diameter on a short stalk; it looks nothing like a coral. Close inspection, though, reveals lighter-colored polyps scattered over the upper surface of the pad. In a container of seawater, each of the polyps extends eight tiny tentacles. Gardens of sea pansies grow on sandy bottoms close to the beach; at very low tides, patches of these soft purple octocorals are sometimes exposed. A small black-and-white-striped nudibranch, which is a shell-less gastropod similar to a terrestrial slug, is a specialized predator of sea pansies. At night, sea pansies put on a light show of bioluminescence. The chemicals the sea pansy produces to make its light have been extracted by researchers at the University of Georgia for use in research on the biochemistry of vision.

Anemones are large single polyps with a flat disc on their bottoms that they use to hold tight to surfaces. Beachcombers often come across the tan, striped anemone *Calliactis tricolor*, which attaches itself to shells inhabited by hermit crabs. Out of water, the hermit crab anemone is just an elastic, glistening lump. Restored to a tidal pool, the anemone will slowly expand its crown of tentacles, revealing delicate orange, red, or yellow colors. Among fouling-community organisms, one sometimes finds a pale, translucent polyp up to one and a half inches long. This is the ghost anemone, *Diadumene leucolena*.

The bodies of other coastal anemones end in a round digging organ instead of an attachment disc. Several species of these burrowing anemones, which are one to three inches long, tunnel under sandy sediments, expanding their feeding tentacles into water currents just above the surface.

Although similar in appearance to the polyps of corals, bryozoans are not related to jellies. Instead, these peculiar creatures, commonly known as moss animals, are in a separate, and equally ancient, group. Bryozoans, like the jellyfish, have a long evolutionary history. Fossils of calcified bryozoan colonies are common in marine rocks from about 460 million years ago right up to the present.

The individual members of a bryozoan colony are tiny soft zooids with a cluster of tentacles around the mouth to ensnare passing plankton. The zooids are housed in gelatinous or calcified structures called zooecia. The body of each zooid consists of one large stomach and an opening near the mouth through which any undigested food is carefully eliminated so that the poop particles aren't recaptured by the tentacles.

Bryozoans, a conspicuous part of fouling communities, are often found growing on oyster shells and dead sea whips. According to the online "A Guide to Benthic Invertebrates and Cryptic Fishes of Gray's Reef," over forty species of bryozoans have been described in the hard-bottom community that lies a few miles east of Sapelo Island. These species often live up to their common name, moss animals; they do resemble moss, although they are usually gray or brown rather than green. Pilings and floating docks are home to the ambiguous bryozoan *Anguinella palmata*, which forms soft, mud-colored, densely branching colonies. The tufted bryozoan, in the genus *Bugula*, grows on shelly and sandy bottoms; after storms it can be found in abundance at the high-tide line. Close inspection of a *Bugula* colony will show the zooecia growing in a double spiral on the branches. Lacy crusts, species in the genus *Membranipora*, are another common type of moss animal encountered on the Georgia coast, appearing as white encrusting colonies with box-like zooecia. For a time, I was intrigued by springy gray masses that sometimes completely covered the black cores of dead sea whips. After exhausting all possible types of sponges, which I had supposed the mass to be, I finally found the mysterious beast in the bryozoan section of a guidebook. It was the rubbery species called dead man's fingers, *Alcyonidium hauffi*.

Although barnacles live in shells made of calcium carbonate plates attached to hard surfaces, they are crustaceans, cousins of crabs and shrimp. Underwater, a barnacle's feathery appendages, called cirri, extend out from its shell and comb the water for food. At low tide, the barnacle closes its plates tightly with a trapdoor in order to keep from drying out. The most common coastal barnacles are acorn barnacles, species of *Balanus*, which can be found attached almost anywhere in the intertidal zone. Some acorn barnacle relatives are more specialized. The small gray barnacle *Chthamalus fragilis* lives in high-tide zones within the estuary. The gray barnacle is distinguished by the tough brown membrane that it uses to stick to surfaces, rather than the white lime plate of acorn barnacles. The shells of horseshoe crabs and blue crabs are often home to the crab barnacle, *Chelonibia patula*, which occurs nowhere else. A related barnacle species is found only on sea turtle shells.

Goose barnacles, in the genus *Lepas*, are very different from other barnacles. These crustaceans live in the open ocean, attached to planks and other floating objects, and are occasionally seen washed up on the beach after a storm. The body of the goose barnacle is encased in hard white plates on a thick bluish-purple stalk. In the Middle Ages, these barnacles were thought to be the young of a wild bird, the barnacle goose, because the coloration of the goose's head and neck bears a striking resemblance to the color pattern of goose barnacles.

Looking more like sponges than anything else, tunicates, or sea squirts, are actually the spineless animals most closely related to fish and other vertebrates. The larval tunicate has a hard rod in its tail, the notochord, which is the evolutionary precursor of the vertebrate skeleton. The common name derives from the spurts of water produced by the two siphons of the adult tunicate when the body is squeezed. Sea grapes, in the genus *Molgula*, are a ubiquitous part of fouling communities in Georgia estuaries. These tunicates occur as clusters of smooth pearly gray spherical bodies, one half inch or more in size, often covered with debris.

Wandering down the beach once after a winter storm, I picked up what seemed to be a chunk of pink rubber. On closer inspection, I decided it must be some sort of animal, but could not imagine what. Its slick, knobby surface was decorated with lighter-colored star-shaped markings. Such a distinctive creature could not be hard to identify, and I was right, though I never would have guessed that the lumpy thing would turn out to be a colonial tunicate. It was a sea pork, *Aplidium stellatum*, which lives on the bottom in subtidal waters; the star-shaped marks were the closed mouths of the individual tunicates embedded in the firm matrix of the colony.

Sound Swimmers *Nekton*

Stars of nature films featuring life in the open sea are the big swimmers: sharks, tuna, and whales. Scientists who study plankton ruefully term these animals "charismatic megafauna." Although they excite the most public interest, they aren't nearly as significant to the daily operation of marine ecosystems as the tiny algae and zooplankton that support the big guys (why this is so is explained in chapter 2). Aquatic animals strong enough to make headway against a current are termed "nekton," from a Greek word that, unsurprisingly, means "swimming." Many species of nekton—for instance, larger shrimp, swimming crabs, and fish—start out as larvae so small that they are part of the plankton. The larvae soon grow big enough on a diet of diatoms and copepods to navigate the sea. Along the Georgia coast and in the estuarine sounds, these strong swimmers include invertebrates such as shrimp and squid and many kinds of vertebrates, including a huge variety of fish, from small minnows to giant tarpon, and mammals such as lively bottlenose dolphins and whales.

Squid

When I was a little girl, my family went to see the latest Disney movie playing at the cinema in Daytona Beach, Florida, where I grew up. The film was *20,000 Leagues under the Sea*. I confess that when the giant squid attacked the *Nautilus*, I hid behind the seat in front of me. That squid was terrifying. Perhaps that is why I eventually became a microbial ecologist dealing with organisms that don't have suckers the size of dinner plates, rather than a teuthologist, a student of squids.

Aerial view of Doboy Sound and the Atlantic Ocean beyond. Myriad swimming animals inhabit these coastal waters. The eight dark humps at the edge of the marsh on the other side of the sound are shrubs growing on heaps of rock ballast left by nineteenth-century sailing ships.

It is hard to believe that fast, brainy squid are kin to oysters and snails. But they are indeed mollusks, in the molluscan class Cephalopoda, which means "head-foot." These active, big-eyed creatures, which include squid and octopi, have highly developed nervous systems. The camera eyes of cephalopods are superficially constructed like our eyes, but in some ways are even better. The light-sensing nerves in their eyes lie behind the retina, whereas in mammalian eyes the nerves are on top of the retina, resulting in a center blind spot. Cephalopods are also able to sense light polarization, which we cannot do without polarizing lenses.

Relatives of these molluscan mariners have been around a long time. There were big-eyed, tentacled nautiloids squirting amid, and perhaps dining on, all those jellyfish in the ocean world of 450 million years ago. These squid ancestors left abundant fossils of their straight or spiraled shells in rock formations. The beautifully weird chambered nautilus of present-day tropical seas gives a sense of the spiral-shelled ammonites that maneuvered though Ordovician seas. Other modern cephalopods don't have a shell, although cuttlefish have an internal calcium carbonate plate that stiffens their soft bodies—and gives parakeets something to nibble.

There are only soft-bodied squid swimming in Georgia estuaries; most common are the long-finned squid, *Doryteuthis pealeii*, and the brief thumbstall squid, *Lolliguncula brevis*, smaller and with more rounded fins. A cephalopod relative, the Atlantic octopus, *Octopus vulgaris*, which can have an arm span of two to three feet, occurs in deeper offshore habitats. Cephalopods have green blood (like the Vulcans of *Star Trek*). Oxygen is carried to their cells by large copper-containing molecules. Unlike the red, iron-rich hemoglobin of vertebrates, the copper-containing molecules are not very efficient at their job, so active cephalopods such as squid can survive only in well-oxygenated water. For this reason, squid avoid estuarine waters poor in oxygen, and quickly die if put in a closed aquarium. Both squid and octopi have large cells in their skin capable of rapid color change. In small squid larvae, these cells, the chromatophores, are fascinating to watch, especially when using a hand lens or low-power microscope. In a matter of seconds, the chromatophores expand and contract in a pattern of winking dark and light spots.

Decapods

This group of crustaceans, whose name means "ten-footed," includes free-swimming crabs and shrimp that are part of the nekton. The chitinous exoskeleton of arthropods, in contrast to the internal skeleton of fish, reptiles, birds, and mammals, works well only for relatively small animals. There is a limit beyond which the external armor becomes too heavy to be operated by the creature's muscles. Despite what 1950-ish science fiction movies would have you believe, ten-foot-tall ants and bus-sized grasshoppers aren't possible. This size limit is less strict in water than on land, since buoyancy offsets some of the exoskeleton weight. But even in the sea there are no really massive arthropods; king crabs and lobsters are among the largest. The most abundant crustaceans of the subtidal estuary are the tiny calanoid copepods in the zooplankton. Decapod crustaceans are also significant contributors to the food webs of coastal systems. Marsh fiddler crabs and grass shrimp are prime food for other wildlife, and the blue crabs and penaeid shrimp in the estuary and near-shore habitats are prime food for us.

Penaeid shrimp include valuable commercial species. The white shrimp, *Litopenaeus setiferus*, grows up to eight inches in length and is translucent with black specks. The dagger-like rostrum extending over the eyestalks—a continuation of the carapace—has short grooves running along either side. The brown shrimp, *Farfantepenaeus aztecus*, is similar in size, but brownish orange in color, and it has longer grooves along the rostrum, extending to the back of the carapace. A third species, the rock shrimp, *Sicyonia brevirostrus*, has a short rostrum and a hard, ridged exoskeleton over the abdomen. The shells of cooked rock shrimp are harder to remove than those of other penaeids, but their flesh is prized for its sweet, lobster-like taste. All three shrimp species are most active at night, feeding on detritus and small animals on the bottom; during the day, they burrow into mud or sand. Adult shrimp live offshore. The females lay their eggs on the seabed, and after hatching, the little larval shrimp ride tidal currents to the shelter of the estuary, where they spend the first part of their lives. As they mature, the young shrimp gradually move to feeding grounds offshore.

While many coastal crabs, like those in the salt marsh, are bottom dwellers, some roam freely in the water and over the seabed, feeding on plant material or on live and dead animals. The most abundant of these free-swimming crabs is the blue crab, *Callinectes sapidus*, which supports a productive fishing industry along the Atlantic and Gulf Coasts. Blue crabs can easily be captured in tidal creeks with a crab trap or a baited line and net from May to October. Soft-shelled crabs are recently molted crabs whose new armor has not yet hardened.

In spring and summer, female blue crabs brood egg masses beneath the broad apron on their lower carapace. (In contrast, the apron of male blue crabs is narrow, shaped something like a lighthouse.) Each egg mass, or sponge, may contain a million or more developing larvae. As the larvae grow bigger within the eggs, the sponge gradually becomes darker, changing from the yellow color of freshly deposited eggs to orange and, finally, to brown when the eggs are ready to hatch.

Like other crab larvae, blue crab hatchlings spend several weeks in the plankton, changing size and shape with each molt until at last they resemble miniature adults, only about an eighth of an inch long, and settle onto the bottom. The tiny, immature crabs overwinter while buried in the soft estuarine sediments. The following spring, they emerge to feed and grow quickly into mature, breeding adults. The females mate once, usually during late summer or early fall. Male crabs can deposit packets of sperm under the female's apron only right after the she-crab has molted, when her shell is soft. To ensure that the timing is right, mating crabs remain locked together in a tight embrace for several days before the female molts, and then for a couple of days after to allow her new carapace to harden. After hibernating in the bottom of the estuary during the winter, the female crabs use the stored sperm to fertilize one or two batches of eggs the next summer. In the fall, two-and-a-half-year-old she-crabs move out to sea, where most of them will not survive another winter. The males rarely live much longer, up to four years at most. In these short life spans, blue crabs grow prodigiously, from microscopic larvae to adults weighing half a pound and inhabiting carapaces up to nine inches across.

The salt marsh is a major feeding ground for blue crabs. At high tide, blue crabs migrate with the incoming water far into the marsh, where they capture killifish, periwinkle snails, marsh crabs, ribbed mussels, and other animals. The crabs also feast on succulent cordgrass roots. One fall, a student sampling the crab population in a tidal creek caught blue crabs whose guts were tightly packed with bits of *Spartina* rhizome. When he later went out to the marsh and carefully picked and washed off some grass roots to taste for himself, he discovered that they were very sweet. When cordgrass stops growing in late summer, it stores up sugars in the roots and rhizomes for new growth the next spring. Blue crabs obviously knew about this long before marsh scientists came along.

Two other free-swimming crabs live mainly in sandy bottom habitats. The hard upper shells, the carapaces, of these crabs are frequently found washed up on the beach. The legs and other body parts of dead crabs are usually nibbled and chewed away by predators and scavengers, leaving only the carapace to be found by beachcombers. The speckled crab, *Arenaeus cribrarius*, is similar in shape to the blue crab, but smaller and mottled with pinkish-white spots on a brown background. These crabs can be seen in the surf; their shells are often found on the sand. The lady crab, *Ovalipes ocellatus*, is freckled with oval, leopard-spot-like markings; its carapace is about as wide as it is broad, with five small spikes on each side and three spikes over the eyes. Lady crabs live on subtidal sandy bottoms; their carapaces are common in beach debris.

Fish

On July 30, 1985, two men from Macon, Georgia, and a local guide were trying their luck with rod and reel in a small outboard in Doboy Sound near Sapelo Island. They got the surprise of their lives. One of the men hooked a medium-size sea trout. At that very moment a huge tarpon, which had been chasing the same fish, seized the trout. The fisherman instinctively hauled back on the line, which, with the forward momentum of the lunging tarpon, carried the monster right into the little boat. The men took refuge as best they could on the motor while the six-foot-long tarpon flailed about, wreaking havoc on the boat's seats and gear for a good half hour. The fish was finally subdued; it was verified as weighing over 142 pounds, a record tarpon for Georgia. This is one of the more spectacular fish stories of the coast, but everyone has a favorite. Mine is of the two-day visitor to Sapelo Island who decided to spend a couple of hours surf casting on the beach. He came back with a twenty-five-pound red drum caught on a six-pound test line with a metal lure.

Fish are abundant along the Georgia coast. They range in size from tiny anchovies to large sharks. Commercially caught fish include shad, whiting, red snapper, and grouper. Sport fishing is an even bigger industry than its commercial counterpart. Besides tarpon, fish sought by local anglers include sea trout, black sea bass, red and black drum, flounder, and sheepshead. Mullet, sea catfish, and croaker are also caught in large numbers in these waters. Estuaries and salt marsh creeks are vital nurseries for the young fish of most of these species as well as for some of the offshore sports fish.

When we lived on Sapelo, our border collie, Yofi, loved going to the beach. When you said, "Go beach?" she cocked her head and perked up her ears. Most particularly, she loved herding the schools of tiny fishes that congregated in the shallow sloughs on the beach at low tide. People on the beach who saw Yofi running and jumping around the sloughs would ask whether the dog was a bit crazy. We would explain that she was expressing her champion sheep dog bloodline as best she could. She also herded sand flies and ghost crabs.

One day our family went to the beach with a small seine to catch some of the slough fish. After a couple of minutes of dragging the seine through the few inches of water left in one of the sloughs, we pulled up about a dozen four-inch-long baby mullet. Our two little boys went wild over the haul and mangled several of the fish before we were able

The author's husband, Barry, getting ready to toss a tennis ball to our border collie, Yofi, in the calm surf of Nannygoat Beach, March 1982. Yofi also loved to herd little fish in the beach sloughs.

to persuade them to gently return the catch to the water. Several more passes netted us a small school of juvenile silver perch, shiny fish with yellow fins, and silversides, nearly translucent little fish with a broad silver band from head to tail. We also caught a couple of striped killifish, gray with distinctive black stripes down the sides. Yofi, pursuing some of the fish we returned to the slough, was shocked when she managed to catch a dazed mullet. She mouthed it and dropped it. A few minutes later, we found that most of the hermit crab population in the slough was fighting over the carcass of the little fish.

The fish we netted in the beach slough were typical of the smallest fish in the estuary: a mixture of the young of larger coastal fish and of species of fish that never grow to more than a few inches long. The smallest fish, which occur in teeming multitudes in the inshore waters, are vital as food for larger animals in the coastal food web. Among these are the silversides, *Membras martinica* and species of

Menidia. Only two to three inches long, these beautiful little fish, as mentioned, bear an iridescent silver stripe along both sides of the otherwise translucent body. Silversides, which abound in the surf and other shallow coastal waters, are food for shorebirds and bigger fish. The bay anchovy, *Anchoa mitchilli*, is a similar little fish found throughout the estuary. It also has a silver stripe along the body, but unlike silversides, the bay anchovy has a deep, laterally compressed body as well as only one dorsal fin. One can quickly tell a bay anchovy from a silverside by bending the fish head to tail. The anchovy's flattened body can be bent nearly double, while the silverside's chunky body will resist being folded.

During the summer, young menhaden, *Brevoortia tyrannus* (known to coastal residents as pogies), school along the coast and in tidal rivers. From a distance, one of these menhaden schools resembles a dark, restless blob on the surface of the water, something out of the imagination of Stephen King. As the blob approaches, the dark patch resolves into hundreds of two- to four-inch-long menhaden swimming in tight formation, filtering plankton from the water as they move. Menhaden are a highly valuable resource for coastal birds, fish, and dolphins as well as for people. In his book *The Most Important Fish in the Sea: Menhaden and America*, H. Bruce Franklin writes: "From the 1860's to the present, catching menhaden has been far and away the nation's largest fishery." The common names "menhaden" and "pogy" are both corruptions of Native American words for fertilizer; this abundant, easily caught estuarine fish was the species that Squanto advised the Pilgrims to plant along with their corn seeds. Schools of larger menhaden are harvested offshore for their high oil and protein content; the fish are used to make omega-3 fatty acid supplements and food for pets and livestock, as well as for bait for commercial and recreational fishing. Pogies are also netted by locals as bait for catching larger fish.

Another valuable commercial fish along the Georgia coast is not a year-round resident, but instead comes once a year to inshore waters to breed. In this respect, the American shad, *Alosa sapidissima*, is like wild salmon. Shad live in the offshore ocean and migrate to coastal rivers to spawn from January to March. During this migration, shad are gillnetted by local fishermen for income during the shrimping off-season. I fondly remember going shad fishing with my father and brother during the winter in Florida: you could throw in an artificial lure and hook a two- to three-pound fish, complete with roe, in a few minutes. On the Georgia coast in late winter, smoked shad is a delicious addition to the usual fare in local fish stores.

Kingfish, or whiting, genus *Menticirrhus*, are the second most important commercial fish species. These panfish are silvery gray, with small, fine scales, a slender body, and two dorsal fins, the first one higher than the second. There are three species of kingfish in southeastern coastal waters, but even trained ichthyologists have trouble telling them apart.

The most prized game fish in Georgia is the tarpon, *Megalops atlanticus*. The last ray of the tarpon's dorsal fin is markedly elongate, and tarpon scales are very large. Adult tarpon do not usually jump into boats. They are caught on poles by surf fishers and by boat in deep places, "tarpon holes," in Georgia sounds. Young tarpon frequent tidal creeks, feeding on killifish and grass shrimp. The ladyfish, or ten-pounder, *Elops saurus*, a smaller relative of the tarpon, is generally found in the same habitats as juvenile tarpon.

Most of the coastal sport fish belong to the Sciaenidae, or drum, family of fishes. This large group of active, carnivorous fish includes sea trout, spot, croaker, and kingfish as well as several types of drum. The young of all of these fish depend on the salt marshes for food and shelter until they grow big enough to avoid predators in the open estuary. Two species of abundant small sciaenid fish are the silver perch, *Bairdiella chrysoura*, a silvery fish with a rounded yellowish tail, and the spot, *Leiostomus xanthurus*, a small deep-bodied fish several inches long, with a prominent dark shoulder spot.

The premium fish in Georgia estuaries for catching and eating is the spotted sea trout, *Cynoscion nebulosus*. Although sea trout are landed year-round, in the fall they congregate in tidal rivers in great numbers. When the first cold snap hit the coast, we knew that it was time for serious trout fishing. At slack tide, we would head for the docks after supper, our poles baited with live shrimp, or if we hadn't gone cast netting that afternoon, with plastic salty dogs. When the sea trout were biting, we could hook several good-sized fish in a short time. The weakfish, or summer trout, *Cynoscion regalis*, grows larger than its spotted cousin. The common name of the weakfish refers to its "weak" mouth, in which hooks are hard to set.

The Atlantic croaker, *Micropogonias undulatus*, is also caught all year long. This fish is trout-like, but has a deeper body, wavy dark bars running down its sides, and a large dark spot on the pectoral fin. A popular activity for marine ecology classes on field trips to Sapelo Island was to suspend a watertight microphone from the dock at Marsh Landing. In addition to the background popping of snapping shrimp, the students were always surprised by the piglike grunts that emanated from under the dock. These, of course, were croakers living up to their name, making reverberating sounds by contracting their air bladders, for reasons best known to themselves.

The largest sport fish in the sciaenid group are the drums. Red drum, *Sciaenops ocellatus*, have reddish bodies two feet or more in length. Red drum are often caught in beach surf, especially during the fall. Local residents of the coast engage in a more challenging method of fishing for red drum: marsh stalking. Fishing pole in hand, we would wade into the salt marsh during a late afternoon high tide, looking for the emerging tailfins of large drum as they hunted fiddler crabs in the shallow water flooding the marsh. When the approximate course of one of the big drums was determined, a hook baited with shrimp or a chunk of mullet was gently landed just in front of the fish's path. The drum quickly found the bait, but patience was needed to be sure the bait was swallowed before pulling the line tight. Then the drum gave a hard, furious fight among the cordgrass stalks. The thick broiled fish steaks enjoyed later in the evening capped off a successful marsh stalk.

The black drum, *Pogonias cromis*, is a related large fish caught in the lower estuary. Adults of this species are dark colored and have numerous hanging barbels (which look like fleshy whiskers) on the chin. Smaller drums, usually less than six inches long, include the star drum, *Stellifer lanceolatus*, a trout-like fish silvery in color, and the banded drum, *Larimus fasciatus*, characterized by its large mouth and seven to nine black bars running down the side.

During an exhaustive, three-year study of fish caught in trawls in Georgia estuaries, a group of researchers found that the four most common fish, making up about two-thirds of the total haul, were sciaenids: spot, croaker, star drum, and weakfish. Half the fish caught during their study were netted in tidal creeks; this result dramatically highlights the role of salt marshes in nurturing estuarine fish populations.

A number of species of fish not related to drums are also caught on the Georgia coast. The sheepshead, *Archosargus probatocephalus*, is a large black-and-white-striped fish with unusual broad teeth that it uses to crop sessile animals off pilings. The sheepshead is a popular and tasty fish, but tricky to catch because those sharp teeth can bite through fishing line. One day, while waiting at Marsh Landing for the Sapelo ferry to depart for the mainland, I noticed two grandmotherly women in a Boston whaler tied up to the dock. One woman carefully baited a hook with a sand fiddler crab, which the other one was too squeamish to touch. The squeamish lady then lowered the baited hook into the water next to a piling. In less than a minute, she yanked the line and brought up a foot-long sheepshead. The first lady efficiently removed the hook from the fish and added a new fiddler to it. Two minutes later, another good-sized sheepshead was landed. The team got one more fish before the ferry departed. I wondered what their total catch was that day. They surely had the secret touch for catching sheepshead.

Other sport fish include young black sea bass, *Centropristis striata*; its body is dark brown or gray, and the fins are large and dark. Striped bass, *Morone saxatilis*, is the basis of a big sport fishery in brackish parts of Chesapeake Bay. In saltier Georgia estuaries, striped bass are caught mostly in the tidal freshwater channels of coastal plain rivers. Florida pompano, *Trachinotus carolinus*, are distinctive, silvery fish with oval, laterally flattened bodies, narrow dorsal and anal fins, and deeply forked tails; their scales are very small and circular. Young pompano are common off the beach from early spring until late fall.

Flounder are a prized food fish, and in Georgia are sold commercially as well as angled for pleasure. Flounder are flatfish; unlike other fish, their bodies are flattened from top to bottom, and the eyes are located on the topside. This body plan perfectly adapts these fish to life on the bottom. They can quickly hide from predators by lying on the sediment and using their fins to cover themselves with sand or mud. They are also chameleons of the fish world, blending into the bottom by changing the color and pattern of the skin beneath their scales to mimic the surface on which they happen to be resting. The summer flounder, *Paralichthys dentatus*, and the southern flounder, *Paralichthys lethostigma*, grow up to a foot or more in length.

Flounder can be caught from a boat by baiting a hook with shrimp and lowering it to the bottom. But a more efficient way to fish for flounder is night gigging. This requires some equipment: an underwater spotlight and a three-pronged gigging spear. One summer a group of students doing research at the Marine Institute built a floating box that held a car battery hooked up to a powerful spotlight. Going out to the beach at night during a low tide, they became quite proficient at finding flounder resting on the sand surface in the calm, shallow water next to shore, and often returned with a dozen or more fish.

The hogchoker, *Trinectes maculatus*, is a flatfish in the sole family, which means that its eyes are on the right side of the upper body rather than on the left side, which is the case for flounders. The common name is rumored to come from observing pigs feast on fish discarded as bycatch (pigs will try to eat anything); apparently they had trouble swallowing this small sole because of its hard, rough scales. Hogchokers are abundant in the estuary, even in freshwater habitats. Blackcheek tonguefish, *Symphurus plagiusa*, are more elongated than other types of flatfish, with a stubby tail that blends into the short fins present around the edge of the body. The pale body of the tonguefish has variable dark markings and a distinct black patch on the cheek. Like flounder, tonguefish have eyes on the left side of their bodies; and like hogchokers, they have no pectoral fins.

Catfish live all over the estuary. They are a nuisance to anglers because they so readily gobble up baited hooks intended for other fish. The two common species are the gafftopsail catfish, *Bagre marinus*, and its cousin the sea catfish, *Ariopsis felis*, which has four instead of two barbels on the lower jaw. (These "cat whiskers" give the fish its common name.) Barbel number is a useful diagnostic when bringing up catfish on a fishing line; the gafftopsail (two jaw whiskers) makes better-tasting fried catfish than does the four-whiskered sea catfish. These fish don't have scales; their skin is covered in a slimy mucus for protection. Catfish also have nasty serrated spines that stiffen their side and back fins, which can cause a wicked cut. I once stepped on a mummified catfish carcass buried in sand on the beach, and discovered from the sharp pain in my foot that dead fish aren't necessarily innocuous.

A fish much underappreciated for food is the striped mullet, *Mugil cephalus*. Abundant everywhere in southeastern coastal waters, mullet have broad heads and grow up to a foot in length. If you notice many fish jumping and splashing in shallow water, they are likely to be mullet. Because mullet scavenge on the bottom for algae and plant detritus, they are seldom caught by hook and line. But as if to make up for this deficiency, it is hard to throw a cast net in the estuary without catching one. Fresh mullet, split open and smoked over live oak coals, is a true coastal delicacy.

Many other species of fish occur in Georgia estuaries, but usually go unnoticed unless accidentally hooked or netted. Some of these fish are attractive, some unusual, and others spectacularly ugly. The species that wins the ugly contest fins-down is the oyster toadfish, *Opsanus tau*. A squat, monstrous-looking fish several inches long with an enlarged head and mottled, wrinkled skin, the toadfish lives around docks and oyster reefs. Toadfish can be an unpleasant surprise catch when fishing around such sites. Other fishy monsters in the estuary are the sea robins, species of *Prionotus*. These medium-size fish, less than a foot long, are distinctive because the side, or pectoral, fins are large and winglike, with three of the fin rays detached for use as feelers. Sea robins have big bony heads and large mouths. Because they sit on the bottom, supporting themselves with their side fins, they are often brought up in estuarine trawls.

Some coastal fish, in contrast, are quite lovely. The Atlantic spadefish, *Chaetodipterus faber*, has a flattened, circular body with five dark vertical bars from the head to the base of the tail. Spadefish range all over the estuary, even far up into tidal creeks. The Atlantic bumper, *Chloroscombrus chrysurus*, has an oval body ending in a narrow, deeply forked tail. Bumpers are common in the estuary during the summer. One of the most distinctive estuarine fish is the lookdown, *Selene vomer*; its eyes are placed high above the long snout on its silvery, stout body. Lookdowns resemble young pompano and, like them, are found in shallow waters of the beach and estuary during the warm months.

Other fish on the Georgia coast are long and slender bodied. The American eel, *Anguilla rostrata*, is easily recognized by its long, snakelike form. Adult eels spend most of their lives in freshwater, but then travel far out into the Atlantic Ocean, beyond the Gulf Stream, to spawn in the Sargasso Sea. The larval eels live as plankton in the open ocean for over a year. When they are strong-enough swimmers, they move to the coast and migrate into freshwater habitats. Here they spend the next five to twenty years before returning to the open sea to mate and die. Eels are often found in tidal and freshwater creeks along the Georgia coast in summer.

Pipefish, in the genus *Syngnathus*, have long thin bodies encased in bony rings, with a small mouth at the end of an elongated snout. Several species of pipefish are found in the estuary, where they feed on small zooplankton. These fish are close relatives of the lined seahorse, *Hippocampus erectus*, which lives in hard-bottom habitats offshore. The northern needlefish, *Strongylura marina*, is slender like the pipefish but has a soft scaly body and long, pointed jaws with numerous small sharp teeth. The darting forms of needlefish are often seen at the water surface in the estuary, especially around docks.

Unlike bony fish, rays and sharks belong to a more ancient line of fish that never developed a hard, bony skeleton. Instead, the bodies of these sea creatures are supported by a soft skeleton made of cartilage, the same material that gives form to our ears and pads our knee joints. In place of scales, they have rough, sandpapery skin. Stingrays, in the family Dasyatidae, have a flat, disc-shaped body with a sharp spine on the top of the whiplike tail; they range all over Georgia estuaries and near-shore ocean. Several species of sharks also live here. Most attain a length of six to fourteen feet, and are often caught by fishermen surf casting on ocean beaches. The most common sharks are the sand tiger, *Carcharias taurus*; sandbar shark, *Carcharhinus plumbeus*; finetooth shark, *Carcharhinus isodon*; lemon shark, *Negaprion brevirostris*; and hammerhead shark, species of *Sphyrna*. There is a hammerhead shark nursery in the sound off Sapelo Island. During a severe summer drought in 1986, saltwater intruded far into coastal rivers, and stingrays and sharks were reportedly caught twenty-two miles up the Altamaha River.

Sharks are known to follow shrimp boats to feast on trash fish tossed overboard from the net hauls. This fact makes local ocean swimmers nervous, but in fact, authenticated shark attacks on humans along the Georgia coast are extremely rare. Speaking of bites, sharks are born with several sets of sharp triangular teeth, which are constantly shed and replaced during their lifetimes. Some of these castoff teeth, a quarter inch to a half inch long, eventually wash up among drifts of small bivalve shells on the beach. Although the teeth are white when shed, shark teeth found on the beach are tan, gray, or black. The color results from the teeth being buried in oxygen-depleted sediment for a long enough time, usually ten thousand years or more, for the calcium phosphate in the enamel to be exchanged for iron and other minerals; this process preserves the teeth, which otherwise would dissolve in seawater. So the teeth you find are fossils from long-dead sharks that swam the seas before humans beachcombed on the Georgia shore. During our time on Sapelo, the Marine Institute had a display featuring fossil shark teeth two to four inches long, from a past geological epoch when monstrous megalodon sharks hunted in the coastal ocean.

Marine Mammals

The liveliest mammals of the Georgia coast are strictly marine, belonging to the cetaceans, the whale family. Bottlenose dolphins, *Tursiops truncatus*, swim year-round in the estuary and near-shore ocean. Dolphins are curious animals that often follow boats or surface to see what is happening on docks and beaches. On high tides, they travel far up into tidal creeks after fish. There they have an efficient hunting strategy. Snorting and splashing in the shallow creeks, the dolphins intentionally panic the fish into jumping onto the creek banks; then they lunge halfway out of the water to snatch up the stranded prey.

    The short-finned pilot whale, *Globicephala macrorhynchus*, is larger than a dolphin, with a striking black-and-white coloration pattern. Pilot whales travel in large pods off the coast and at times become stranded on beaches. The cause for such stranding remains unclear. The broad continental shelf off the Georgia coast is a nursery ground for right whales, *Eubalaena glacialis*, the rarest of all the large whales. During the era of New England whaling, these whales were the easiest to kill and had a large amount of blubber, so they were known as the "right" whales to hunt. Right whales, including mothers with calves, can be spotted a few miles off the coast during winter. The whales spend their summers filtering out minute plankton in feeding grounds in the North Atlantic.

12 On, and under, the Beach *Living in Sand*

During the years I lived on Sapelo Island, I learned a lot about what lives in the shallow waters off the Georgia coast by wandering along Nannygoat Beach. Beachcombing after winter storms was especially productive. Curious remains of animals that inhabit the near shore were deposited all along the high-tide line. Many of these were easy to identify, but a few were puzzling. One thing was sure: there was a fascinating array of marine life in the sandy world of the surf and beyond. These animals included not only mollusks and crustaceans, but also echinoderms, an ancient group of invertebrates found only in the sea.

Mollusks

The sandy-bottom habitat off the sea beaches is home to a variety of marine snails. Beachcombers delight in finding the shells of the knobbed whelk, *Busycon carica*, which is the official state shell of Georgia. The whelk is easily recognized by the pointed extensions jutting out from the gray, brown-streaked shell. The lightning whelk, *Busycon contrarium*, looks nearly identical to the knobbed whelk, but is a distinct species. When you find a whelk shell, hold it so that the knobbed top is pointed up and the opening of the shell faces you. If the opening is to the right of center, then it is a knobbed whelk; if to the left of center, it is a lightning whelk. The two snails also have different habits. While the knobbed whelk goes prowling in beach sands both day and night, the lightning whelk likes to rest when it is dark and hunts only during the day. Live whelks are gathered by coastal residents for food; the large tough foot is tenderized, chopped into small pieces, and cooked for soups and salads.

The delicate channeled whelk, *Busycotypus canaliculatus*, has a lovely, creamy white shell with smooth deep grooves spiraling to the top. The whelks prey on bivalve mollusks that live in sandy sediments. Whelks lay their eggs in strings of papery disc-shaped capsules often found discarded in beach wrack after the young snails have hatched.

Another gastropod shell loved by beachcombers is that of the moon snail, *Neverita duplicata*, also called a shark eye. The round shell, gray to tan in color and up to three inches in diameter, is a darker color in the center. Live moon snails are seldom encountered because they spend all their time burrowing under the surface of

Delicate, white, channeled whelk (*top*) and robust, striped, knobbed whelk (*bottom*), near-shore snails whose empty shells are commonly found on Georgia beaches and are favorite homes for thinstripe hermit crabs.

On, and under, the Beach

subtidal sandflats, hunting benthic clams that they feed on by rasping a hole in the clam's shell. The underground path of moon snails occasionally takes them near the surface. I discovered this accidentally while trying to figure out who made what tracks on the bottom of a slough in the beach. One strange trail was a rather wide, V-shaped depression that went on for a short distance, ended, then started up again a foot or more away, as if the animal were alternately dragging itself on the sand and then swimming in the water. One of the tracks ended in a circular bump in the sand. Digging with my son's plastic shovel, I unearthed what appeared to be a slimy, half-used bar of soap. It was a live moon snail. Its mantle is so large that it cannot be contracted into the shell, but instead serves as a lubricated plow that the snail uses to burrow through the sand. The mystery was solved; the snail had been traveling underneath the surface of the slough, sometimes emerging to leave a V-shaped wake in the sand. A strange object sometimes found on the beach, a broad sand-and-mucous collar, is also a product of the moon snail, which builds the structure to protect its eggs until they hatch.

Shell of a moon snail. The live animal resembles a white, slimy bar of soap tunneling just under the sand on the lower beach and near-shore bottom. The tapering hole rasped into the shell is the signature of another moon snail that cannibalized its kin.

The baby's ear snail, *Sinum perspectivum*, is a cousin of the moon snail; its flat white shell, up to one and a half inches across, looks like a little ear. Like the moon snail, the baby's ear feeds on bivalves in sandy sediments and has a large slippery mantle. Clay Montague told me that locals call the live snail a "snot dog" because it looks and feels like a big wad of nose goo. The tapering holes, wider at the top, rasped out by these two snails into the shells of their prey are easily distinguished from the straight holes bored by oyster drills. Many moon snail shells found on the beach have such tapering holes drilled into the bottom of the shell, evidence of predation by their kin.

The most beautiful snail shell found on Georgia beaches is that of the lettered olive, *Oliva sayana*. The glossy cylindrical shell, up to two inches long, has dark triangular markings on a yellow-brown background. This gastropod is a deadly predator of bivalves living along the shore. Olives feed at night, burrowing through the sand in search of small smooth-shelled clams that they kill by smothering them with their muscular foot. The Atlantic auger snail, *Terebra dislocata*, also has a reputation for preying on small clams in the sand. The auger's narrowly tapering shell is up to about two inches long, with many distinct ribs along the whorls.

Unlike other gastropod mollusks, the slipper shells, *Crepidula fornicata* and *Crepidula plana*, are not spiraled, but are flat or concave with a shelf across part of the interior shell, hence the common name. Slipper shells are also not like other gastropods, since rather than living as mobile predators, they spend their lives attached to pilings or to other shells, and feed by filtering phytoplankton and small zooplankton from the water. Slipper shells, like many other sessile creatures, like to lodge on the snail shells appropriated by hermit crabs. The dome-shaped *Crepidula fornicata* settles on the outside of the hermit's shell, and the flat *Crepidula plana* often finds a secure place on the inside entrance of the shell.

Bivalve mollusks living in the intertidal beach and just offshore are more diverse, producing many interesting shells for beachgoers to discover. Two small bivalves occur in great abundance in the surf zone: the little surf clam and the coquina. The small white shells of the surf clam, *Mulinia lateralis*, are usually less than an inch long and somewhat triangular in shape. Even smaller, the shell of the coquina, *Donax variabilis*, is more elongated and, true to its species name, quite variable in color. Coquina shells can be pale yellow, pink, or purple, and are often striped. Opened up, the shell halves resemble a delicate butterfly. In Florida, coquina shells are so abundant that there are rock formations along the coast composed mainly of these little clams. The Spanish fort of Castillo de San Marcos in St. Augustine was built of stones mined from local coquina rock, which is soft and easy to quarry, but hardens over time with exposure to air.

Both the surf clam and the coquina burrow into the sand of the lower beach and then use long siphons poking out of the surface to filter phytoplankton from the surf. When a strong wave erodes the sand from the little clams, exposing them to sandpipers, they all quickly dive back down into the beach as fast as their tiny feet can take them. Another small clam, only a quarter inch long, *Abra aequalis*, prefers the less turbulent subtidal sediments.

Other near-shore bivalves are much bigger. The one- to two-inch-wide disk clam, *Dosinia discus*, has flat, glossy white, circular shells scored with fine ridges. Disc shells often bear a small hole that a snail rasped out to feed on the soft body within. The robust Atlantic cockle, *Dinocardium robustum*, is even larger, up to five inches across; opened up, the two ribbed shells have the shape of a heart. The cockle's shell is yellowish outside with brown markings, and a lovely pink inside.

Like the cockle, the two arks found on Georgia beaches have shells that are thick and ribbed, but more oval than those of the cockle, and with a distinctive flat ridge with little teeth where the two shell halves join. The blood ark, *Lunarca ovalis*, is tinted orange red and can be two inches long. This species has red hemoglobin in its blood (or hemolymph) to bind and carry oxygen to the tissues, just as we, and all vertebrate animals, do. The ponderous ark, *Eontia ponderosa*, somewhat larger than the blood ark, has a white shell covered with a fuzzy black coating that is usually worn away in a "bald spot" on the top.

Some beach bivalve shells are so fragile that an intact specimen is a prized find. A paper-thin white oval shell up to three inches long, somewhat lopsided with concentric ribs, belongs to a channeled duck clam, *Raeta plicatella*. A truly beautiful bivalve shell is that of the angel wing, *Cyrtopleura costata*, which constructs pure white, wing-shaped shells with scaly ribs. While the angel wing can grow to seven inches in length, its smaller relative, the false angel wing, *Petricolaria pholadiformis*, is at most two inches, with fewer ribs along the shell. Despite their delicate shells, both angel wings are champion burrowers. The angel wing bores holes up to three feet deep in subtidal sands. The false angel wing prefers muddy sediments and will sometimes drill into submerged wood. The tellins, in the genera *Tellina* and *Macoma*, are thin-shelled clams that burrow less deeply. The largest of these, *Tellina alternata*, has a shiny white oval shell several inches in length, tinged with yellow at the edges.

Some bivalve shells are long and narrow. The stout tagelus, *Tagelus plebeius*, has a chalky white oblong shell three times as long, up to four inches, as it is wide. The jackknife clam, or razor clam, *Ensis directus*, makes a thin gray-blue shell six times longer than it is wide. The largest bivalve found along the Georgia coast is the saw-toothed pen shell, *Atrina serrata*; the thin, triangular brown shell with long serrated ridges running down it can reach up to a foot in length. These large bivalves live in the coastal ocean off the beach, anchored in the sand by tough byssal threads.

The bivalve species that inhabit the sandy beach and near shore feed on plankton in the water or on microbe-rich detritus that accumulates on the bottom. Some of these mollusks—for example, pen shells—are sedentary, embedded in the sandy seafloor in one location. Others, like tagelus and razor clams, live in deep burrows in sand or mud, coming to the surface to feed and then moving down into their holes to escape predators. Many beach bivalves, in contrast, are as mobile as snails. Cockles actively move about, burrowing rapidly through the sand or propelling themselves over the bottom with thrusts of their muscular feet. The tellins are especially rapid burrowers and often change location in the sediment. The thin shells of the tellins make them more buoyant than heavier-shelled clams, enabling them to stay in the flocculent layer of mud and detritus at the sediment surface.

Crustaceans

Like other coastal habitats, the seashore has its complement of crustaceans adapted to living on the beach or in the waters lapping the sands. These range from the small, entertaining beach hoppers and ghost crabs to the mysterious shrimp burrowing in the intertidal sand to the large marine crabs that roam in the near-shore ocean and whose carapaces decorate the wrack left by high tides.

In beach sands live numerous amphipods less than a half inch long, species in the genera *Haustorius* and *Acanthohaustorius* and their relatives. Amphipods are crustaceans whose bodies are compressed, or flattened, from side to side. Like harpacticoid copepods, amphipods are a ubiquitous, diverse group of benthic animals. The haustoriid amphipods are a staple in the diet of sandpipers, which incessantly probe the moist sands of the intertidal beach.

Other amphipods include beach fleas, *Orchestia grillus* and *Americorchestia longicornis*, which inhabit wrack deposits on the upper beach. *Orchestia grillus* can also be found underneath dead plant litter in the salt marsh. Beach fleas are most active at night, when the sand is cool and they are less likely to be spotted by birds looking for a meal. Many amphipods hop about on their tails and rear legs. The especially athletic beach fleas can leap many times their body length, just as insect fleas do.

At low tide, the beach is pocked with myriad quarter-inch-wide holes. Where the sand is still wet, small geysers spout from some of the holes. All over the beach, tiny, dark brown fecal pellets litter the sand, encircling the holes and piling up in drifts by the surf. One professor who often led class field trips to Sapelo Island liked to play a joke on his students. While lecturing on beach life, he would show them a handful of the fecal pellets and then pop them into his mouth with relish. The "pellets" he ate were actually chocolate sprinkles, which look exactly like the cylindrical brown fecal matter. The subterranean animals responsible for producing such great quantities of "chocolate sprinkle" pellets are ghost shrimp, *Callichirus major*. Ghost shrimp, along with related burrowing crustaceans found from the intertidal beach to the bottom of the estuarine sounds, are called shrimp because of their elongated body shape, but they are, in fact, more closely related to lobsters and crabs.

The ghost shrimp is a strange-looking beast, up to four inches long, with an elongated abdomen and large, slender front claws. These animals spend their adult lives in two- to four-foot-deep holes in the sand, constantly pumping water through the burrow for food and oxygen. Because they lead a sheltered existence, ghost shrimp exoskeletons

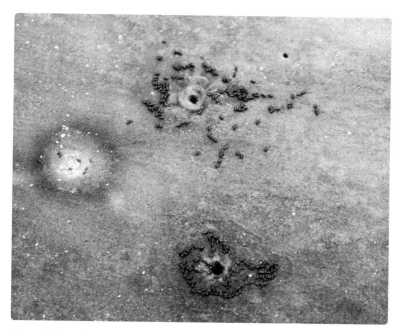

Ghost shrimp burrows strewn with brown fecal pellets from plankton meals. The shrimp feces are the exact size and shape of chocolate sprinkles, which a professor used to fool his field trip students into thinking he was eating.

are soft and flexible compared to those of water-dwelling shrimp and crabs. Their burrows are narrow at the top, opening up into a roomier living space about an inch across. When runnels on the beach erode away the top few inches of ghost shrimps' holes, the larger bottom sections appear as "chimneys" jutting up from the sand, held intact by their builders' mucous secretions.

Because a ghost shrimp's burrow is so deep, the little creature is nearly impossible to dig up. I have tried on numerous occasions, attacking a hole with quick thrusts of a shovel. Invariably, the ghost shrimp vanished into the depths as the excavation collapsed the walls of its tunnel. Other creatures find refuge in the security of *Callichirus* burrows. A common burrow mate is a tiny commensal crab, *Austinixa cristata*, a close relative of the pea crab found in oysters. A second species of ghost shrimp, *Biffarius biformis*, is smaller than the beach shrimp and lives in deeper parts of the subtidal estuary, where water turbulence is not as great as on the beach.

While fiddler crabs and blue crabs are true crabs with well-armored bodies, anomuran crabs must contend with a soft, shell-less abdomen projecting behind the head and legs. The best-known members of this group are hermit crabs, the clowns of the crustacean world. These cousins of true crabs protect their vulnerable abdomens from predators by occupying empty snail shells. As the crabs grow larger, they must find successively bigger shells to protect their tender nether parts. Sometimes suitable shells are in short supply. The shells cannot have holes drilled into them by moon snails, for instance, or the hermit's abdomen could be subject to the sharp poke of another hermit crab's claw through the hole. If an empty shell is not available, a needy crab will attempt to evict another hermit from its shell. Usually this tactic is unsuccessful, but the sparring matches that ensue are amusing to watch.

The largest hermit crab along the Georgia coast is the thinstripe hermit, *Clibanarius vittatus*. This monster gets big enough to drag around knobbed whelk shells. I found three of these hermits in a crab trap once, attracted by the rotten fish heads in the bait well; all were in hefty whelk shells. Other species of hermit crabs in this region are much smaller, with plain white claws. Two abundant ones are the dwarf, or common, hermit crab, *Pagurus longicarpus*, and the flat-clawed hermit crab, *Pagurus pollicaris*. At times, great aggregations of these little hermits can be found in beach sloughs, scavenging for detritus and dead animals washed up by the waves. These *Pagurus* crabs sport a wide variety of shells, from tiny spiral auger shells on very young crabs to two-inch-wide moon snail shells on the largest adult hermits. Sometimes the crabs will have shells of mudflat snails, even though the nearest living mud snails are on intertidal marsh flats over a mile away from the beach. It is likely that empty mud snail shells are carried to the beach by the tides.

Another anomuran crab bears little resemblance to true crabs. Mole crabs, *Emerita talpoida*, are queer little creatures with streamlined oval bodies one half to one inch long, abundant in the surf zone of high-energy beaches along the Atlantic coast. But since the surf on Georgia beaches is not strong enough to suit mole crabs, they are a rarity here.

The common semiterrestrial shore crab is the ghost crab, *Ocypode quadrata*, a close relative of fiddler crabs. These attractive crabs have a boxy sand-colored carapace up to two inches across and yellow-tinted claws. Their color and nocturnal behavior make them aptly named. Ghost crabs excavate deep tunnels in the upper beach and sand dunes; larger crabs live farthest from the sea. They spend the daylight hours renovating their burrows or resting deep in the cool, moist sand. After sunset, they emerge to spend the evening feeding on the beach. Seen silently scurrying over the sand in the moonlight, they really do seem ghostly. Besides scavenging dead animals washed up by the surf, ghost crabs hunt live prey such as snails and clams in the surf zone. Like other land-dwelling crabs, ghost crabs release their broods into the ocean to live as plankton. The final, megalopa stage of ghost crab larvae has a strong rounded carapace that protects it from the surge of the waves as it makes its way into its new home on the beach.

The shells of other true crabs that live just offshore are often discovered washed up on the beach, especially after storms or extreme tides. The most strikingly marked of these coastal crabs is the Dolly Varden, or calico, crab, *Hepatus epheliticus*. The broad carapace of this crab is tapered toward the back, up to two and a half inches in length, and ornamented with irregular dark-bordered red spots on a light orange-brown background. The calico crab conceals itself just under the surface of subtidal sandflats, where it preys on mollusks and hermit crabs. The smaller purse crab, *Persephona mediterranea*, has a light-colored hemispherical carapace with darker splotches and three evenly spaced spikes at the rear margin. Like the calico crab, the purse crab hides in subtidal sand for protection and as cover from which it ambushes its prey, usually small crustaceans and fish.

The light brown shell of the spider crab is teardrop shaped, up to four inches long, and covered with spines or bumps. The long legs give the crabs a decidedly spidery look. The two common species are nearly identical, differing mainly in the number of spines down the back of the carapace—about nine for *Libinia emarginata* and six for *Libinia dubia*. Spider crabs roam over sandy bottoms, feeding on small bivalves and crustaceans. They are also known to feed on starfish by grabbing one of the spiny arms. The starfish drops off the captured arm to escape (it will later grow another), and the crab then turns its prize around to feast on the soft inner tissues. The bumpy surface of a spider crab carapace makes a good attachment site for fouling animals. Spider crabs are often so overgrown with sponges, hydroids, and bryozoans that they are also called decorator crabs.

The tiniest coastal crabs live in association with other animals. The pea crab, *Zaops ostreus*, is a parasite living within the shells of oysters, where it nibbles off parts of the oyster's gills. A little pink pea crab can often be found inside the shell of a roasted oyster. The quarter-inch-wide crab *Dissodactylus mellitae* inhabits the underside of keyhole urchins. This tiny crab is a true commensal animal. It uses its host only for protection and for free meals as it catches the crumbs of food left over from the urchin's dinner; but it does not harm the urchin. The urchin crab is found only on sand dollars and keyhole urchins. Other small commensal crabs, species in the genus *Pinnixa*, live in bristle worm burrows in the subtidal estuary. These crabs find refuge in the back entrance of the polychaete's home, filtering particles from the water currents pumped through the burrow by the worm.

Horseshoe Crabs

The largest crustacean shell one is likely to come across on the beach is that of the bizarre-looking horseshoe crab. The horseshoe crab, *Limulus polyphemus*, is not a true crustacean; it is more closely related to spiders (arachnids). This unmistakable arthropod has a large brown chitinous carapace up to a foot across—which in outline does resemble the shape of a horseshoe—and a long spiky tail jutting behind. Underneath, five pairs of dark brown jointed legs dangle over leathery book gills. The gills are often parasitized by a flatworm, the limulus leech, *Bdelloura candida*.

Horseshoe crabs are living fossils; remains of crabs essentially identical to modern species have been found in rock layers dating to the Ordovician period, 450 million years ago. Today, several species of horseshoe crab still roam the shallow waters of seacoasts just as they did before the era of the dinosaurs, along the North American Atlantic and Gulf Coasts and in the western Pacific Ocean along the coasts of Japan, China, and Indonesia.

These curious crabs scuttle about on subtidal sand and mud bottoms, feeding on small animals in the sediment. In the spring, adult horseshoes migrate up onto the beach to mate and lay batches of eggs in moist sand, where the developing larvae will be sheltered from deeper-water predators until the baby crabs hatch and return to the sea. Often, breeding horseshoe crabs become disoriented on the beach and can't find their way back to the surf. Their dead bodies provide a feast for ghost crabs.

Since 1970, the blood of horseshoe crabs has been used to test for the toxins of bacteria that cause human disease. White blood cells of this ancient species release a chemical that binds to bacterial toxins, causing the blood to coagulate. This is probably a natural protection against bacterial infection. During the spring mating season, horseshoe crabs are collected as they come ashore to lay their eggs. A small quantity of blood is extracted from each crab by inserting a syringe needle into the base of the tail, then the crabs are released. The white blood cells are treated so that they lyse and release the bacteria-binding compound.

Echinoderms

"Echinoderm" means "spiny skin," which certainly is true of the starfish, sea urchins, and sea cucumbers in this phylum. Echinoderms have a body divided into five parts around a central core, either a calcareous skeleton or calcium carbonate spicules in the skin, and a multitude of little tube feet on which they move about the sea floor. The familiar members of this group are the starfish. The common sea star,

Asterias forbesi, is orange brown with a red-orange spot, a light-sensitive organ called the madreporite, on the upper surface. The arms of the sea star extend out to several inches. The margined sea star, *Astropecten articulatus*, is about the same size, but has a pale purple body outlined by a brown ribbed margin. The common sea star attacks bivalves, using the brute strength of the tube suction of its arms to force the two shells apart; it then everts its stomach through its mouth to digest the unlucky clam. The margined sea star prefers preying on small gastropods, tunneling through the sand to find a victim, then engulfing the whole snail with the star's extensible mouth.

Brittle stars, abundant in the estuary, have a small central body disc surrounded by five fragile, whiplike arms. The spiny brittle star, *Ophiothrix angulata*, lives in intertidal muddy sandflats and around oyster reefs. Unlike the predatory sea stars, the fast-moving spiny brittle star is a nocturnal deposit feeder, scavenging for particles of food on the bottom. The most common brittle star in the estuary, *Hemipholis elongata*, anchors itself to the bottom by burying two or three of its long arms in the sand or mud and then extending its other arms into the tidal current to capture plankton floating by.

Every beach visitor recognizes the keyhole urchin, *Mellita quinquiesperforata*. The long species name simply means "five perforations," referring to the five open slits, or "keyholes," in the urchin's shell. The flat, disc-shaped skeleton resembles that of its northern cousin, the sand dollar. The live animal is greenish brown and can be found feeding on detritus in moist sand at low tide. The circular outline made by the urchin is easy to spot in the wet sand. As the urchin burrows just under the surface of the beach, sand and water is pushed up through the keyhole slits. A clever scientist, Malcolm Telford, figured out how this helps the urchin remain buried. In a 1981 paper in the *Bulletin of Marine Science*, Telford described how he used dyes to track the flow of water across a burrowing urchin. Water flowing up and through the holes in the urchin's shell relieves water pressure that builds up on the underside of the flat shell, thus keeping the urchin from floating up out of the sand.

A common echinoderm of the Georgia coast is the hairy sea cucumber, *Sclerodactyla briareus*. The dark fat cylindrical body, several inches long, has rough, hard knobs all over the surface. The more fragile worm sea cucumber, *Leptosynapta inhaerens*, has a slender white body. Inhabitants of subtidal sediments, sea cucumbers feed on plankton and detritus from within shallow burrows. The hairy sea cucumber is sometimes washed up after storms, its rough elastic body drawn up into a ball and covered with shell fragments. Picking up one of these shelly balls, one is hard-pressed to imagine the poor beached sea cucumber as a relative of sea stars and sand dollars.

13 Loggerheads

The loggerhead turtle, *Caretta caretta*, is a true marine reptile. English mariners gave the turtle its common name; they thought that the thick head of the turtle resembled a log. Loggerheads are found in tropical and semitropical seas all over the world. The hard, dark green shells of these huge sea reptiles are two to three feet long; adult turtles weigh as much as a large man, one hundred to four hundred pounds. Only soft-shelled leatherback sea turtles grow bigger than loggerheads. The turtles are long lived, with a life span of fifty to seventy years. They are slow to reach maturity, becoming capable of breeding only after they are twenty to thirty years old.

Normally, the big turtles range far out into the open sea, feeding on a multitude of species of crabs, squid, fish, and other sea creatures. Loggerheads can dive deep, up to 750 feet, in search of clams, sponges, sea urchins, and anemones growing on the bottom. In beds of sea grass, divers have observed a loggerhead digging a shallow trench in the sediment to find tasty morsels, advancing along its path by shoving sand away from its face with its front flippers. As clams and worms burrowing in the sand are exposed, they are snapped up in the turtle's beak. Loggerheads also rest on the bottom for hours, safely out of the way of larger predators and surface motors. But to reproduce, they are still tied to their ancestral life on land.

During my time on the island, I was fortunate to watch a female loggerhead in the ancient turtle ritual of laying eggs in a nest on the upper beach. On a fine June evening during a full-moon high tide, a group of us walked quietly up Nannygoat Beach. We knew that the loggerheads mate in spring when the females hook up with the males hanging out in turtle migration paths in the open ocean. While the females are dedicated to their home beach, returning faithfully to lay their eggs in the sand in which they were hatched, the males are roamers who mate with females headed in any direction. This way, the turtle population keeps its gene pool mixed, meaning that there is a single worldwide species of *Caretta caretta*. Turtles nesting in Australia are genetically the same as the loggerheads that come up on Georgia beaches.

The females take advantage of the highest night tides to assist their travel up the beach. Female turtles can swim a long way to their special nesting beaches. One turtle tagged on a nesting beach on Wassaw Island near Savannah was tracked by satellite over a distance of 900 miles before she was recaptured.

In the moonlight, we spotted a dark shape in the breaking waves, a turtle swimming to shore. We turned off our flashlights and kept well back as the female loggerhead emerged from the surf and laboriously crawled to the first line of sand dunes on the upper beach. After we were sure that the turtle had begun to prepare her nest, we crept closer. A female loggerhead coming onto a beach will turn back to sea if she detects possible danger, but once she begins the egg-laying process, nothing will disturb her dedication to the task. The turtle used her back flippers to dig a deep, cylindrical hole in the moist sand. That done, she started depositing dozens of leathery golf-ball-sized eggs into the nest. The loggerhead's eyes streamed with a constant flow of tears, not from pain but as a natural way to secrete excess salt ingested with the water she swallowed while swimming in the ocean. The salt excretion from special glands in her eye sockets is constant; this allows the turtle to get all the water she needs by drinking seawater and to maintain the salt balance in her blood.

After the turtle finished laying her eggs, she carefully covered her brood with sand and then swept the beach over a wide area with her flippers to conceal the exact location of the nest hole. Finally, the exhausted loggerhead turned away from the dunes and slowly crawled back to the ocean, her flippers scraping the sand on either side with each labored shove of her heavy body. In the morning, conservationists checking the beach would note the distinctive "turtle crawl" tracks made by the loggerhead during her trip to and from the dunes. This turtle would probably lay eggs once more during the summer. But after all of these labors, she would likely take a vacation from breeding for two to three years before repeating the daunting task.

During another evening in late summer, during another high tide, the baby turtles from this and other sea turtle nests would hatch. After digging their way out of the sandy hole, they would all frantically flipper their way to the surf. How many of these babies will be boys is determined not by a male chromosome, but by the temperature of the eggs in which the embryos developed. If the soil surrounding the nest hole was less than 80°F during incubation, most of the hatchlings will be male. If the nest temperature remained at 90°F or more, all will be female. At intermediate temperatures, the eggs at the cooler bottom of the nest will hatch boy turtles, and the eggs in the warmer top, girls. So turtles laying in May will produce mainly male offspring. A return to the beach to deposit another batch of eggs in the warm sand during July will yield more female young. By nesting throughout the summer, the loggerheads are able to maintain a balance of males and females in their population.

After leaving the beach, two-inch-long hatchlings unerringly swim out to sea, cresting each wave rolling into shore. The little turtles bob at the surface; at this stage they are buoyant and can't sink. They swim steadily for as long as a day to escape the beach and near-shore zone where multitudes of birds and fishes with a taste for baby turtle lurk. After their escape, the little ones rest and swim, rest and swim, until they encounter debris or clumps of sargassum weed on the sea surface. There they find shelter and food: little fish, crabs, and jellyfish, which also congregate around any flotsam offering protection from hungry eyes above and below. They live on planktonic animals in the open sea for six to twelve years, gradually getting stronger. They are eventually able to dive down to sea grass beds and rocky bottoms to feed on benthic creatures, as their parents do.

In earlier times, marine turtles were extensively harvested for meat and turtle shells, and turtle nests were plundered for eggs. Protection from commercial use has allowed populations of loggerhead turtles to increase a bit, but other dangers abound. Development of the coastline for recreation has eliminated some former nesting areas. Shrimpers inadvertently net adult loggerheads coming in to lay eggs, although the addition of turtle exclusion devices to shrimp nets has greatly decreased this problem. Once the eggs are laid, raccoons and ghost crabs avidly seek out nest holes filled with these nutritious delicacies. The Georgia Sea Turtle Center on Jekyll Island documents and protects loggerhead nests and operates a hospital for injured turtles.

14 Shorebirds

No seacoast would be complete without flocks of squabbling gulls, terns circling like miniature kites, sandpipers busily poking about after the retreating surf, and pelicans lazily riding the waves. Georgia has an especially rich variety of shorebirds because of the expansive feeding grounds in salt marshes, mudflats, and the open beach, and also because the subtropical climate attracts an abundance of overwintering ducks and sandpipers.

Seagulls are the most conspicuous birds of the beach and open estuary. The two dominant gulls in the region are the gray-backed herring gull, *Larus argentatus*, and its smaller relative, the black-headed laughing gull, *Leucophaeus atricilla*. Other gulls often seen along the coast are the ring-billed gull, *Larus delawarensis*, and Bonaparte's gull, *Chroicocephalus philadelphia*. The stunningly large great black-backed gull, *Larus marinus*, is an occasional winter visitor to the Georgia shore.

Any reasonably sized boat that plies the sounds or coastal ocean has an attendant retinue of herring and laughing gulls hovering astern, sharp eyes on the boat's wake. Now and then a dazed small fish or shrimp will be thrown up by the wash of the propellers, and several gulls will swoop down to snatch the prize. The first bird to grab the tidbit is usually not the winner, since the other gulls will pester him unmercifully until he is sufficiently distraught to release his catch. As the hecklers drop down to fight over the remnants, the first gull dejectedly reassumes his position behind the boat. The crowd of gulls hits the jackpot every time a shrimper stops to shake trash fish from the net haul. Every gull within view of the boat swiftly joins the turbulent mass of shrieking birds fighting over the scraps.

Terns are easily distinguished from gulls by their slimmer bodies, narrow and sharply pointed wings, and forked tails. Terns and gulls differ also in temperament. Terns tend to be solitary, aloof fishers. The smaller species hover in place for minutes over a likely spot, then in an eyeblink dive in headfirst to snap up an unwary fish. Most terns seen along the Georgia coast are light bodied with gray wings and a black cap. Among the usual crew are the small least tern, *Sternula albifrons*, the common tern, *Sterna hirundo*, the Sandwich tern, *Thalasseus sandvicensis*, and the gull-sized royal tern, *Thalasseus maximus*. Least terns have several rookeries on the sea islands, and larger terns are known to nest on Little Egg Island, a part of the Wolf Island National Wildlife Refuge at the mouth of the Altamaha River.

The gull-billed tern, *Gelochelidon nilotica*, is quite different from other terns in appearance and behavior. Looking more like a gull than a tern, the gull-billed is all white, with a black cap in summer, and unlike most terns, is often seen hunting over salt marshes for fiddler crabs and insects rather than plunging into the water for fish.

One of the most striking shorebirds is the black skimmer, *Rynchops niger*. This skimmer has a large white body, a black back and wings, and a large red bill. The black-tipped lower mandible is longer than the upper, for good reason. True to their common name, skimmers fly just above the water along the beach, the lower mandible cutting the surface. When the bird hits a fish, the long beak snaps shut, and the skimmer, still flying, expertly tosses up the catch and swallows it down. Skimmers prefer to hunt at low tide at the edge of the surf, where the waves are reduced to low ripples, or in the line of sloughs left on the beach by the retreating sea. One day a pair of skimmers came flying down the beach when our sons were splashing in one of the sloughs. The birds skimmed the slough just up the beach from us, gracefully skirted around us, and continued skimming farther up. When they got to the end of the line of sloughs, they circled back and repeated the maneuver, again avoiding our little pool. There must not have been many little fishes in the water, since I didn't see them catch anything, but we were treated to a close-up view of these lovely birds.

Another distinctively patterned shorebird is the American oystercatcher, *Haematopus palliatus*. The long red bill, black head and body, and white belly and wing stripe make the oystercatcher easy to spot even at a distance. These birds feed in small flocks on clams, mussels, and oysters. Young oystercatchers learn how to open bivalves from their parents. In the same group of oystercatchers, one sometimes sees several different strategies employed to get at the sweet mollusk meat tightly enclosed in the shell.

Plovers likewise have very striking plumage. Several common species in this region have dark backs, white undersides, white wing stripes, and black-and-white markings on the head. These are the semipalmated plover, *Charadrius semipalmatus*; Wilson's plover, *Charadrius wilsonia*; and the killdeer, *Charadrius vociferus*. The latter's common and species names derive from the loud, sharp call: "kill-deer," given when trying to distract predators from its nest, hidden in the grass. Two related species are the black-bellied plover, *Pluvialis squatarola*, which is light gray in winter, dark chested in summer, and the ruddy turnstone, *Arenaria interpres*, strikingly patterned in white, black, and reddish brown and often seen on oyster reefs or rock jetties.

Everywhere along the shore, sandpipers busily probe into sand and mudflats at low tide for worms, small crustaceans, and other meiofauna. Great flocks of small sandpipers are a common sight on the beach. Many of these little peeps breed on the Arctic tundra and winter along southeastern coastlines. It is hard to distinguish individual species of these tiny brown birds, especially in their drab winter plumage. Among the overwintering species are sanderlings, *Calidris alba*; semipalmated sandpipers, *Calidris pusilla*; least sandpipers, *Calidris minutilla*; western sandpipers, *Calidris mauri*; dunlins, *Calidris alpina pacifica*; and red knots, *Calidris canutus rufa*. Purple sandpipers, *Calidris maritima*, frequent rock jetties.

On high-energy beaches, sandpipers depend on mole crabs and coquina clams that ride the last rush of water onto the beach from the breaking waves, then quickly burrow just under the sand as the wave recedes. These little animals are easy for pipers to catch. On Georgia beaches, mole crabs and coquinas are rare, and the peeps have to work hard to sort quarter-inch-long haustoriid amphipods out from the sand at the surf zone. It is amusing to watch a flock of sandpipers running and probing in unison along the edges of the waves. A mysterious signal will be given, and then they all take off in tight formation to find a more promising spot. The flock flies around in a circular pattern before settling on a new site.

In contrast to other pipers, the spotted sandpiper, *Actitis macularius*, is a solitary bird whose summer plumage has distinctive brown spots. It is easily recognized by its habit of continually bobbing its tail up and down. The female spotted sandpiper, unlike most birds, has it made. After she mates with a male, she lays her eggs and leaves him to brood and rear the chicks. The female then seeks out a second male, with whom she mates again, only to abandon him with another clutch of eggs.

Larger sandpipers generally hunt alone or in small groups along marsh creek banks and on the intertidal beach. The greater and lesser yellowlegs, *Tringa melanoleuca* and *Tringa flavipes*, are grayish pipers with bright yellow legs. They are hard to tell apart except for the difference in size. Short-billed dowitchers, *Limnodromus griseus*, have dull, light gray plumage in winter but turn reddish in summer, with a long white rump patch. Whimbrels, *Numenius phaeopus*, are a species of curlew; these medium-size speckled brown birds have white-and-black-striped faces and long, dark, down-curved bills.

The willet, *Tringa semipalmata*, is recognized by its black-and-white wing pattern in flight and loud "willet-willet" call. Willets seem to be just as fond of feeding on marsh mudflats as in the surf on the beach, and often nest in the high marsh. When I went out into the marsh during summer, invariably I got too close to a willet nest. Two distraught birds would suddenly be swooping over my head, plaintively calling my attention away from the ground. I never found a willet nest, but I always knew when I moved the magic distance away from it, since the nervous parents would abruptly disappear. Two distinct subpopulations of willets frequent the Georgia coast. One group overwinters in the tropics and migrates to the southeastern United States to breed. The willets seen here in winter favor northern marshes for raising their young and then take over the beach and mudflats when the summer willets head south in the fall.

Large sandpipers were once popular game for bird hunters. In the late nineteenth and early twentieth centuries, many a fashionable restaurant featured gourmet meals with willet or curlew. Now shorebirds are protected, but only after many species were brought to the edge of extinction. The common names of large pipers often derive from the hunting era. Yellowlegs, for instance, are also called tattlers because these high-strung birds would be the first to raise a noisy alarm when shooters were spotted.

In contrast to the sandpipers, the brown pelican, *Pelecanus occidentalis*, rarely seems to go to any great exertion to find a meal. Pelicans are usually observed either bobbing about in the middle of the sounds, stolidly standing in small groups on spits of sand with their long heavy beaks resting on their breasts as they survey the waves, or leisurely flying in a loose line just above the waves. Only occasionally have I seen a pelican swoop down with a great splash to scoop up a fish in its commodious pouch. These birds must enjoy such great fishing success, and get such large fish each time they go hunting, that they can laze around most of the day. In Georgia, brown pelicans nest in a few sites, notably on Egg Island Bar and Satilla River Marsh Island.

Other diving birds in the estuary appear to lead equally laid-back lives. Double-crested cormorants, *Phalacrocorax auritus*, largish jet-black birds with long necks, are generally spotted perched on the tops of pilings and waterway markers. Immature cormorants can be told from the adults by their light-colored breasts. In freshwater parts of the estuaries, one may find a black cormorant-like bird with an even longer neck sitting with wings outstretched on a log. This is the anhinga, bearing the appropriate scientific label *Anhingha anhinga*. Also dubbed the water turkey or snakebird, anhingas dive completely underwater for fish, as do cormorants. Because these underwater swimmers don't have waterproof feathers, both cormorants and anhingas spend a great deal of time between dives drying out their soaked wings in the sun.

One must search carefully to see another diving bird common in the estuary. Horned grebes, *Podiceps auritus*, are small dark birds with two little tufts of feathers sticking out over their ears. The grebes sit low in the water and are quick to dive under when disturbed. Often I thought I had spotted the head of a grebe among the waves, but by the time I raised my binoculars for confirmation, it had disappeared; now you see it, now you don't. If I kept scanning the water for a minute or so, I would see the little tufted head pop up again some distance away.

During winter, Georgia estuaries resound with the calls of migratory waterfowl, specifically ducks and their relatives, which visit in incredible numbers. A scientist visiting the Marine Institute, who happened to be an avid small-game hunter, couldn't believe the abundance of wild ducks in the marshes around Sapelo Island. I couldn't believe his accurate identification of the birds. Walking with him at the edge of a *Spartina* marsh, we flushed out a couple of ducks from a tidal creek. All I saw was a dark blur of wings as the ducks shot up from the water and streaked for the open sound. But the visitor was excited. "Black ducks," he exclaimed immediately, "they are hard to shoot." He told me he identified them from "the way they flew."

I was able to identify only sitting ducks, but fortunately there are plenty of those from October to March. The most numerous are lesser scaup, *Aythya affinis*. These dark-headed light-backed ducks sit in flocks of thousands in the open sounds. During the day, they disperse to hunt for food, but at night they return to huddle in masses in the protected waters of the estuary. At times the scaup flocks were so thick during the winter that it was hazardous to run a small boat across Doboy Sound at night. Other ducks that take up winter residence in the estuaries include mallard, goldeneyes, redheads, American wigeon, canvasbacks, and buffleheads. Teal, ring-necked ducks, gadwalls, shovelers, pintail, wood ducks, and ruddy ducks are more often found in freshwater ponds along the coast. Flocks of scoters, together with occasional eiders and old-squaws, rest in the waves just off the beaches, and red-breasted mergansers can be seen here and there in the sounds.

My hands-down favorite duck is the hooded merganser, *Lophodytes cucullatus*. Several pairs of these birds spent the cold months in the pond by the laboratory, where at least in winter they were safe from hibernating alligators. Male hooded mergansers are spectacularly handsome, with a black-edged white crown and a pure white chest offsetting the dark body. The females, in the best avian tradition, are drab brown and gray, but do sport an elegant golden crest. One cold, bright winter day, I noticed a lone duck paddling over the salt marsh at high tide, disappearing for half a minute at a time and then bobbing up a few feet farther along. After running for my binoculars, I discovered it was a male hooded merganser, diving to feed on small invertebrates on the marsh surface.

There is an official shooting season for ducks along the Georgia coast, for a week in late November and then in December and January. But those who prefer to hunt with binoculars can enjoy the ducks all winter long.

Many resident shorebirds nest along the Georgia coast. Laughing gulls hide their eggs in the upper cordgrass marsh. These gulls, like the clapper rail, have completely adapted to raising young in the tidal marsh, since their eggs can tolerate short periods of inundation at high tide. Herring gulls, terns, and skimmers lay eggs in depressions in the sand in gregarious colonies, favoring open dunes and unvegetated areas of small islands. Dredge spoil deposited on the salt marsh, raising the level above flood tide, can form an ideal nesting site. The seabird colonies are sensitive to human disturbance; when nesting colonies are visited too frequently, they are usually abandoned and then relocated in a more private spot.

Seasons in the Sun   15

It is a beautiful, clear, calm day in January. The sun shines on the sparkling water of the sound and warms the air. The estuary and marshes are eerily quiet, as if holding their collective breath before spring arrives. The seasonal cycle is ready to begin anew. All of life in this teeming, productive complex of mud and grass, tidal creek and open sound, is cued to great and small rhythms of nature: seasonal variations in day length and temperature, monthly cycles of spring and neap tides and moonlight, and daily tides and alternate light and darkness. There are also less predictable events: winter storms that bring bouts of wind, intense summer thunderstorms, and occasional autumn hurricanes.

On October 2, 1898, the strongest hurricane to hit the Georgia coast made landfall just south of Sapelo, on Cumberland Island. The storm surge in Brunswick was recorded at sixteen feet. The 1898 hurricane is part of the story tradition of the resident community on Sapelo, passed down from one generation to another. The brunt of the winds and high water fell full force on the island. Much of the land was underwater, and houses were washed away. People climbed the oak trees and tied themselves down to keep from drowning. At that time, there was a hospital on Blackbeard Island, just north of Sapelo, to quarantine and treat crews of ships infected with yellow fever, a cause of serious epidemics in the late 1800s. The hospital was completely destroyed in the hurricane. But not all was lost, as Cornelia Bailey relates in the book *God, Dr. Buzzard, and the Bolito Man*, a first-person account of growing up on Sapelo Island. She tells how the big storm aided the construction of a community church among a cluster of houses known as Raccoon Bluff on the north end of Sapelo, facing Blackbeard Island: "High winds tore the hospital apart and sent the lumber from it floating over toward the Bluff. The people at the Bluff asked permission to get the lumber and then they rowed out and gathered it up and they built themselves a new church. And that's how the church came to be located at the Bluff."

In September 1989, another major hurricane took aim at the Georgia coast. As it neared, Hurricane Hugo was projected to make landfall close to Sapelo. Our family was still living on the island when all the residents were ordered to evacuate. We could take only one suitcase per person and our pets. While we hurriedly packed, we thought about tales of the 1898 hurricane. Riding the ferry over to Meridian Dock, under lowering clouds blown by gusting winds that raised whitecaps in Doboy Sound, we envisioned our house and laboratory swamped by surging floodwaters. We loaded our two youngsters, Aaron and Jared, and our border collie, Yofi, into our mainland station wagon, parked in a garage at the dock, and headed north.

Since we figured the inland motels would be full of coastal refugees, we decided to drive up to Atlanta to stay with my parents. We saw convoys of Georgia Power trucks headed south in advance of the storm. As the day darkened and we gained Highway 75 north, my husband, Barry, tuned the car radio to an Atlanta station. After a hurricane update, droning sitar music by Ravi Shankar came on. A couple of minutes into the piece, I noticed Barry's head starting to droop. He was falling asleep at the wheel! Sitting in the back with our younger son, Jared, I yelled at Aaron in the front to poke his dad awake. He did, and Barry snapped up. A radio station with livelier music was located, and we arrived safely at my folks' house. Luckily for the Georgia coast, Hugo swerved north as it approached, and instead made landfall near Charleston, South Carolina. Other than a few downed trees, the buildings on Sapelo and the beaches and marshes around the island, escaped serious harm.

While hurricanes are rare events, life in the estuary must still cope with everyday environmental stresses. Coastal plants and animals need to be very adaptable. They must survive changes in salinity in the open estuary and tidal creeks, which ranges from nearly freshwater after a winter storm or a summer shower to ocean saltiness during droughts. The salt content of coastal seawater fluctuates from near 0 to 35 parts per thousand, the equivalent of 3.5 grams of salt per liter, or about half a teaspoon dissolved in a quart of water. (This is also the amount of salt, 3.4 grams, that the average American consumes each day.) Because of the evaporation of saltwater in the upper marsh, cordgrass, saltwort, and other high-marsh plants grow in sediments with salt contents that reach 60 parts per thousand, nearly twice that of ocean water.

Water temperature extremes within the estuary are also severe, ranging from near freezing during winter cold snaps to over 100°F in tidal creek pools at low tide under the summer sun. In addition, plants and animals living in the estuary must withstand strong tidal currents, with maximal water velocities approaching four knots; unstable, shifting bottom muds and sands; and murky, sediment-laden water. It is a tribute to the resourcefulness of nature that so many species are able to thrive under these challenging conditions.

The six- to nine-foot tidal range along the coast shapes the animal and plant communities found here. Twice a day, the salt marshes become part of the sea at flood tide, and then an extension of the land when the tide ebbs. The plants that have managed to colonize this intertidal habitat have done so at great metabolic cost. Cordgrass and other marsh plants must be able to grow in oxygen-poor, waterlogged muds, and at the same time be able to withstand periods of inundation, when their leaf pores are tightly shut and they cannot carry on photosynthesis. At low tide, the marsh plants endure desiccation because of the high salt content of the marsh soils. In addition to these paradoxical conditions of too much and too little water, noxious chemicals in the soils, mainly sulfide, the compound responsible for the rotten-egg smell of estuarine mud, need to be detoxified to prevent root death.

Cordgrass, *Spartina*, is well adapted for growing in this harsh marsh habitat. This plant, like other grasses, has a special type of photosynthesis, called the C-4 pathway, in which the first organic sugars made from carbon dioxide and water have a backbone of four carbon atoms.

Most other land plants, such as trees and herbs, use a more common pathway called c-3 photosynthesis, in which the first sugars have three carbon atoms. Plants with the c-4 pathway are more efficient at capturing carbon dioxide and are better able to function at high air temperatures than plants using the c-3 pathway.

The tough, fibrous stems and leaves of *Spartina* resist animal grazers, except for squareback crabs, marsh grasshoppers, and leafhoppers (or periwinkle snails during one of their occasional population outbreaks), so the plant can allot plenty of the organic matter it produces to new growth. The hollow stems and underground rhizomes afford a pathway for the flow of oxygen from the air down to the roots and out into the surrounding soil. The oxygen combines with reduced iron in the sediment to produce iron oxides. When a mass of cordgrass roots is dug up, channels of red running along the plant roots mark where oxygen carried down from above has "rusted" the iron-rich mud.

The element sulfur is necessary to make proteins. It was thought that cordgrass might get sulfur by absorbing it from the abundant sulfate in tidal water when the marsh floods. But an elegant test showed that the roots take up sulfide from the soil instead. This bit of information was discovered by using a geochemical technique known as stable isotope analysis. Scientists measured the relative amounts of the two stable (nondecaying, so nonradioactive) forms of the element sulfur: s-32 and s-34. The s-34 form is a bit heavier than s-32 because of two extra neutrons in the nucleus of the atom. Seawater sulfate has more of the heavier sulfur isotope than does hydrogen sulfide in marsh muds. The relative amounts of s-32 and s-34 in *Spartina* leaves would tell where the plant got its sulfur. Researchers were surprised that the isotope ratio of organic sulfur in marsh cordgrass matched the sulfur isotope fingerprint of soil sulfides rather than that of seawater sulfate. *Spartina* plants were utilizing the toxic sulfide rather than the more easily absorbed seawater sulfate. It is likely that oxygen degassing from the roots allows sulfide-oxidizing bacteria to convert the poisonous sulfide to the nontoxic sulfate form before the roots take up sulfur from the soil.

Spartina expends a great deal of energy in regulating the salt content of its tissues. The plants concentrate sodium chloride in order to produce an internal saltiness close to that of tidal water; this helps prevent water loss and wilting. Still, quantities of freshwater must be extracted from the marsh soils to compensate for the water lost to evaporation through leaf pores during photosynthesis. The roots can exclude salt, but some is inevitably absorbed. Excess salt is secreted from special salt glands in cordgrass leaves. Naturally, plants growing farther back in the marsh, where the soil is saltier and the tidal washing of salt off the leaves is less frequent, are more salt stressed than plants lining tidal creeks.

The metabolic cost to marsh plants of salt stress has been analyzed by carefully monitoring gas exchanges at the leaf surface of *Spartina* and other plants growing in low- to high-marsh sites. During photosynthesis, carbon dioxide is taken out of the air and oxygen is released. When, in the course of keeping itself alive, the plant burns up some of the sugars it has made, in the metabolic process known as respiration, carbon dioxide is given off. The balance between oxygen produced and carbon dioxide released tells scientists what portion of organic carbon made by the plant is used up in respiration. The smaller the proportion of organic carbon burned up during metabolic respiration, the more left to grow more plant biomass. Respiration by *Spartina* growing along tidal creeks was found to consume only about 11 percent of the plants' total production of organic carbon. In contrast, plants in the flat marsh plain beyond the creeks used up 45 percent of the sugars produced daily during photosynthesis simply to survive. Plants growing in the salt flats at the landward edge of the marsh showed even greater metabolic costs. For glasswort, saltwort, and sea oxeye, 65 percent to 95 percent of the organic carbon they produced was burned up in respiration to keep the plants alive, leaving very little for new growth. These metabolic results nicely mirror what is evident when one visually surveys variation in plant growth: tall, luxuriant stands of creekside cordgrass, shorter cordgrass in the marsh plain, and sparse, stunted herbs at the marsh edge.

Most animals in the marsh synchronize their activities to the rhythm of the tides. During low tide, fiddler crabs swarm over the marsh surface, scooping up sediment and organic detritus, scraping off algae and microbes from particles, and then depositing whatever is left in neat little balls of mud. At high tide, the fiddlers retreat into the shelter of their burrows. In higher-elevation areas of the marsh, fiddlers plug up the opening of their burrows with firm mud or sand. Sometimes they plug their burrow entrance on extreme low tides as well, to keep their holes from drying out in the hot summer sun. As the tide creeps in, lower-marsh fiddlers retreat into their burrows, but plug the entrances only during very high tides. Fiddler crabs have internalized the tidal clock to such an extent that in a laboratory, with no external cues, they become active during the local time of low tide, and quiescent at the time of high tide. Animal behaviorists have found fiddler crabs to be an excellent model to use in studies of the biochemical basis of endogenous rhythms.

During low tide, other animals more adapted to life on land than in the sea carry out their daily routines in the marsh. Wharf crabs scurry about, marsh snails rasp over the marsh surface, insects suck and chew on cordgrass leaves, marsh birds hunt the insects, and raccoons prowl for fiddler crabs and mussels.

With the incoming tide, animals native to the sea are released from their refuges in tightly closed shells or in tidal creek pools. Oysters, ribbed mussels, and clams open their valves and poke siphons into the flooding water. Even though their feeding time in the marsh is limited to high tide, the mussels process an incredible amount of suspended material when water floods the marsh. During summer, most of the deposition of sediment onto the marsh surface comes from the mud and particles expelled in the feces of filter-feeding mussels. Mud crabs, which can't breathe in air, emerge from their watery holes to feed. Killifish and grass shrimp swim far and wide over the marsh, busily consuming small invertebrates. Larger fish and blue crabs range through the cordgrass stalks, seeking snails and errant fiddler crabs that have not yet retreated into their burrows.

Some animals remain active over the entire tidal cycle. Periwinkle snails and mud snails can feed underwater as well as when the marsh is exposed. Herons and egrets stalk fish among the cordgrass stalks at high tide and along the edges of tidal creeks at low tide.

At sunset, the cast of characters changes. Fiddler crabs and seashore birds are mainly active only during the day, although the sand fiddler, *Uca pugilator*, also roams the marsh during low tide at night. Most marsh and dune mammals are nocturnal. Rice rats scuttle in the *Spartina* marsh, and rabbits, cotton rats, and mice come out of hiding to feed in the back dunes. Many beach animals prefer darkness, avoiding the heat of the sand and sharp bird eyes when the sun is out. Ghost crabs and beach fleas scamper among the beach wrack; spiders hunt over the dunes.

Even though the Georgia coast enjoys fairly mild winters, which attract multitudes of northern "snowbird" tourists as well as waterfowl, there is a pronounced seasonal cycle in the marsh and estuary. Winter brings shorter day lengths and increased cloudiness, so marsh plants don't receive enough sunlight to grow. Colder winter temperatures slow the pace of life. Some inhabitants of the coast hibernate in order to avoid lethal freezes that come with cold snaps. Alligators, snakes, and terrapin turtles find shelter in deep holes in the mud or sand.

Other animals adapt to the cold. Periwinkle snails remain active in the marsh, but they stay close to the sediment surface rather than crawling up *Spartina* stalks, as they do in summer. Fiddler crabs retreat to their burrows during cold snaps, but emerge to feed on mild winter days. Subtidal animals are better protected from temperature extremes, since the water temperature changes only slowly. The water and sediments remain warmer than the air most of the winter. Still, in winter the tidal creeks and sounds become much cooler than the open ocean. Many aquatic animals cannot tolerate very cold temperatures, so they migrate out to the warmer shelf waters during winter. Blue crabs, shrimp, and fish that have fed and matured in the estuary during the summer head out to sea in late fall. Smaller estuarine animals that can't manage such a trip simply make the best of it. They hide out in the sediments when the water chills and then resume activity during warm spells.

Winter, however, is a great time for birders. Ducks, sandpipers, and other shorebirds that rear their young as far north as the Arctic flock to ice-free southern estuaries, and many summer resident birds remain on the coast during winter.

Spring comes early to coastal Georgia. Some flowers bloom on the sea islands in February, but life in the marsh and estuary really begins to stir in March with the first balmy weather. Among the initial signs of spring are blooms of dinoflagellates, species of *Kryptoperidinium*, in the tidal creeks. The creek waters become stained red from the masses of this pigmented flagellate, responding to increased day lengths and warmer temperatures. Although a red-tide species, these dinoflagellates are not toxic, as some of their relatives are. At the same time, *Spartina* begins sending up new shoots from its rhizomes, and fiddler crabs emerge from their winter rest. Marsh and estuarine animals mate and produce the first batch of young of the summer season. Soon the coast is buzzing with biting midges and mosquitoes.

The pace of life increases in tempo with longer days and higher temperatures. Curiously, there is a brief pause in midsummer, when marsh plant production slacks off and there is a lull in the abundance of biting insects. Late summer to early fall is the time of peak activity, spurred by the highest air and water temperatures of the year. The marshes and tidal creeks fairly burst with animals feeding, reproducing, and trying to escape predators, right up to the first cold front of autumn.

Marsh Food Webs: Who Eats What, and How We Know That

Although the variety of species in habitats along the Georgia coast is lower than in other marine ecosystems, the sheer number of animals is overwhelming. In one square yard of *Spartina* marsh, one may find up to 150 mud fiddler crabs, 50–300 periwinkle snails, and 8 ribbed mussels. Densities of 1,600 mud snails per square yard are not uncommon on creek banks, and tidal creeks contain millions of small fish and grass shrimp. The question of what sustains the multitude of animals in southeastern estuaries has occupied the minds and research projects of scores of marine ecologists for decades. Work carried out at the University of Georgia Marine Institute on Sapelo Island has been central to this effort.

The original concepts of the food web of salt marsh estuaries were developed by John Teal, Eugene Odum, and colleagues working at the Marine Institute on Sapelo Island. These scientists discovered two things. First, the amount of plant matter produced by *Spartina* in the salt marshes each year rivaled that of the most productive cornfields. Second, very little living *Spartina* was eaten by animals. So they asked

Teal's Boardwalk built over a cordgrass marsh plain at the south end of Sapelo Island, March 1982. This is the salt marsh in which John Teal carried out early research on the role of cordgrass in estuarine food webs.

what happened to all that cordgrass, which died back each fall and regrew in the spring? At the same time, a researcher at the Marine Institute found that algal production in the creeks and rivers was very low because of the murkiness of the sediment-laden water. So *Spartina* must somehow be the main food source for marsh and estuarine animals.

The theory of a detritus-based food web was born from these observations. The basic idea was that dead *Spartina* leaves and stems were broken into smaller fragments by the combined activities of microbes and animals. These small bits of plant matter and their associated microbes, termed detritus, were then used as food by bottom-feeding animals in the marsh and estuary. Predatory crabs, fish, and birds then ate the animals that fed on the detritus. Initial field results supported this view of food webs in the estuary. Creek water draining the marsh was full of pieces of marsh cordgrass. Analysis of material fluxes in the marsh suggested that about half the plant matter produced by *Spartina* each summer was eventually washed out into the estuary by the tides.

But further research threw a wrench into this simple concept of a detrital food web. The marsh and estuarine food web was found to be much more complicated. Careful measurements that accounted for all the detritus in tidal water as it flooded and then receded from the marsh showed that, on average, more detritus was deposited on the marsh surface than was carried away by the tide. Only during rainstorms at low tide, when the marsh plain was exposed, was there a net loss of plant material to the estuary. More sophisticated analysis of production by benthic microalgae on the mudflats and by phytoplankton in the open estuary revealed that algal production was high, and more than enough to sustain a good part of the estuarine food web. This was a meaningful finding because algae are a more nutritious food for coastal animals than are particles of cordgrass, even with the microbes growing on the detritus.

Then, a geochemical study provided new evidence that not all was kosher with the idea of a detritus food web. The method was based on stable isotope analysis, the same technique behind the discovery that *Spartina* takes up sulfur from marsh soils. In this case, sources of organic matter in the estuary were evaluated by the relative amounts of stable isotopes of the element carbon. It turns out that about 1 percent of carbon atoms have an extra neutron in the nucleus, making them one atomic unit heavier, mass 13, than the other 99 percent of carbon atoms, with mass 12. Unlike radioactive carbon atoms of mass 14 (C-14), which are very much less common and subject to decay, carbon-13 atoms are stable—that is, they do not decay. The slightly greater mass of C-13 makes a difference in plant metabolism. The plant enzyme responsible for binding carbon dioxide takes up the lighter C-12 compound slightly more readily than it does the heavier carbon dioxide molecules that happen to have a C-13 atom. As a result, plant organic carbon, made from the carbon dioxide taken in by the discriminating enzymes, contains less C-13 than the ratio of C-13 to C-12 in the carbon dioxide present in air would suggest.

The differences in the relative amounts of C-13 are in fact really, really small, but thanks to the wonders of modern technology, they can be precisely measured. To do this, geochemists use a sophisticated instrument called a double-beam mass spectrometer. This machine compares the isotopic mass ratio of an unknown material with that of a known standard material. The resulting data are reported in parts per thousand (o/oo) difference (del) in isotope content compared to the standard. For stable carbon isotopes, the standard value, written as del 13-C, is set for convenience at 0 o/oo. The standard used for del 13-C spectrometry has traditionally been the calcium carbonate internal shells (something like a cuttlefish's internal shell) of a squid-like cephalopod, *Belemnitella americana*, that swam in the newly forming Atlantic Ocean during the Cretaceous period. There are abundant fossil belemnite shells in Cretaceous limestone rock in the Pee Dee formation of South Carolina. The standard is thus known to geochemists as Pee Dee Belemnite, or PDB. If a sample has relatively more C-13 atoms than the PDB standard, it has a positive del 13-C value. Because plant enzymes don't readily bind to carbon dioxide molecules containing the heavy C-13 atom, the del 13-C values of plant matter are always negative compared to the standard, meaning they have relatively fewer of the heavier c-13 atoms than does the belemnite standard.

A second lucky happenstance in the geochemistry of carbon is that once the heavy c-13 atoms that do get taken up by the plant enzymes become part of large organic molecules, the ratio of c-13 to c-12 in the plant matter becomes fixed in a unique value. Decomposing microbes and animal consumers don't care what the ratios of carbon are in the food they consume. So their tissues will have the distinctive carbon-isotope signature of the plant matter they ate. This isotopic "fingerprint" of a plant source will faithfully transfer up the food web to predators.

Now for the fun part. Fortunately for salt marsh ecologists, all plants do not have the same ratio of c-13 to c-12, or del 13-c value. Grasses with the more efficient pathway of photosynthesis, the c-4 pathway, like *Spartina*, use enzymes to take in carbon dioxide from the air that are different from the ones used by c-3 plants. The cordgrass enzyme isn't as fussy as the c-3 enzyme about differences in the mass of carbon dioxide molecules. A c-4 plant tries to get all the carbon dioxide it can while releasing as little water from the leaves as possible. Some c-13 carbon dioxide is still rejected, though, so its del 13-c value of about -12 o/oo is negative in relation to the PDB standard. This value is still not nearly as negative as the values of c-3 plants, including some upper-marsh herbs and most land plants. The c-3 pathway enzyme really discriminates against the heavier carbon dioxide molecule. The del 13-c values of c-3 pathway plants are -25 o/oo to -30 o/oo. It turns out that marine phytoplankton and benthic algae have a stable-carbon-isotope signature of -18 o/oo to -22 o/oo, right between the values for *Spartina* and most land plants.

So all that scientists needed to do was to take samples of organic matter and animals in the marsh and compare their isotopic fingerprints to those of the marsh plants. The results surprised marsh ecologists. The carbon-isotope ratios of the particulate material suspended in tidal creek water, which had been thought to mainly come from degraded *Spartina* leaves, instead closely matched the del 13-c value

Mass of dead cordgrass stems floating down a tidal creek. Although such rafts are impressive, research has shown that much of the plant matter in tidal creeks originates from algal, rather than cordgrass, production.

of phytoplankton. Only the very largest bits of plant material in the water had the same carbon-isotope fingerprint as *Spartina* plants.

The ecologists were relieved to find out that most marsh invertebrates, including insects, snails, and marsh crabs, had about the same del 13-C value as *Spartina*; thus, cordgrass was the ultimate food resource for these animals. So they weren't wrong about that part of the food web, at any rate. But the filter-feeding mussels in the marsh and the oysters in the creeks had isotope ratios more like those of phytoplankton. Animals in the estuary had a range of isotope values and appeared to be getting as much of their food from algae as from *Spartina* detritus.

The view of food webs in salt marsh estuaries had to be revised. It seems that most of the great annual production of cordgrass is consumed locally, within the marsh and small tidal creeks. Even so, the marshes are still vitally important to estuarine and coastal ocean creatures, particularly as feeding grounds and as nurseries for their young. Instead of being passively carried out of the marsh as plant detritus, *Spartina* carbon swims out in the guts of blue crabs, shrimp, and fish. The new view also includes phytoplankton in the water and benthic algae on the mudflats as a major source of food for coastal wildlife.

16. The Once and Future Coast

Thunk! Whang! The icebreaker shuddered as it muscled through a broad ice sheet several feet thick. Enormous slabs of ice burst up from the cold sea, white on the surface from recent snows, an ethereal blue below. The ice blocks hammered the side of the ship, which resounded with the blows. I trusted that the hardened hull of the U.S. Coast Guard Cutter *Polar Sea* would withstand the battering.

In the summer of 1994, I took part in a historic scientific project to study the central Arctic Ocean by ship. A convoy of two icebreakers, the *Polar Sea* and the Canadian Coast Guard Ship *Louis S. St. Laurent*, was headed for the North Pole, the first North American surface ships to try for that goal. That summer the polar ice cap, from the northern Alaskan coast to the central Arctic, was thick and extensive. Huge pressure ridges formed by the wind-driven collision of large floes reared tens of feet above the ice, and as many feet below. The ridges presented daunting obstacles. To get through them, our ships had to back and ram. This involved first reversing engines, then charging a pressure ridge at high speed, shoving the bow up onto the ridge and knifing down against it. Slicing through a pressure ridge often required several back-and-rams. The ship shuddered each time it rammed the thick ice dam until finally the blockage gave way.

The *Polar Sea* was a study vessel, at that time one of two icebreakers in the U.S. scientific fleet. But in the end, our ship was damaged by the constant ramming and by the impacts of truck-sized ice blocks thumping around and under the hull. A blade of one of the three propellers sheared off, taking a bite out of a blade of another propeller. The *Polar Sea* now had only two serviceable propellers, including the one with the damaged blade. Since we were near the pole, the

The author in front of the Russian icebreaker *Yamal*, at the North Pole, summer 1994. Conditions in the Arctic, especially severe melting of the summer ice pack, may affect climate as far south as the Georgia coast.

Canadian ship broke a path through the ice to our destination. We did our routine sampling, but the rest of our planned cruise looked uncertain. We couldn't return back to Alaska through all that thick ice; the path we had broken had already been closed by the shifting ice pack.

Fortunately, the Russians came to the rescue. By coincidence, one of their nuclear-powered icebreakers, the *Yamal*, was already at the pole, just over the horizon. They had come from Murmansk in the eastern Arctic to film a children's TV show at the North Pole. A helicopter arrived from the *Yamal*. After negotiations among the three captains, our ships were restationed near the Russian ship. All three crews enjoyed ice liberty, with open-house tours of the ships. Despite the fact that the Russians were on an opposite time schedule (our day was their night, so the children were all asleep), the *Yamal*'s crew hosted a barbecue for everyone on the ice, with vodka-laced hot tea and grilled reindeer. (Yes, we ate reindeer at the North Pole.) We took pictures of ourselves posing on an ice floe beside the *Yamal*, a huge red toothy grin painted on the bow. The Russian ship, outfitted for tourism, was certainly more posh than our science vessels. It even had a small swimming pool heated by the nuclear fuel that powered the engines.

The damage to the *Polar Sea*'s propellers meant abandoning our planned course back to Alaska. Instead, we would take the shorter route through the eastern Arctic Ocean, with the more powerful *Yamal* breaking a path through the ice for our ships. At one point, when the *Polar Sea* and *Louis S. St. Laurent* bogged down in thick ice, the *Yamal* steamed in a great circle around us, churning the ice pack to bits so we could proceed. With the Russians' aid, our ships safely arrived off Iceland, where most of the science party disembarked to fly home. Those scientists who unfortunately didn't bring a passport had to stay on the icebreakers to Boston.

My husband, Barry, and I had other arctic adventures after that one. Working in a frozen ocean was quite a change from our research in the steamy Georgia salt marsh estuaries. But as it turns out, what goes on in the Arctic is relevant to the future of the southeastern coastal systems. Scientists term effects in earth systems that act across great distances "teleconnections." These interactions are not well understood, but every year it is more apparent that the state of the northern ice pack is linked to weather patterns that reach to the southern United States.

Since 1994, the ice pack that grows across the Arctic Ocean each winter has become ever smaller during summer. In 1997–98, we took part in a year-long project in which another Canadian ice breaker, *Des Groseilliers*, was intentionally stuck into the ice pack in the Beaufort Sea north of Alaska. Scientists, including Barry, were amazed at how thin the sea ice had become during the summer of 1997. The bottom of the ice floes had melted, taking with it the normal community of ice algae and small animals that served as food for zooplankton and fish swimming underneath.

During the summer of 2004, I was working on another project in the western Arctic Ocean on a new U.S. icebreaker, the Coast Guard Cutter *Healy*. That summer, the ice had almost completely melted away in the Beaufort and Chukchi Seas. The dark ocean water, free of its frosty cap, was several degrees warmer than normal. One day the *Healy* crew saw a polar bear swimming far from land, heading out to sea in search of the vanishing ice pack and its seals. We hoped he was able to make it, though we knew the ice edge was many miles away.

Arctic sea ice continues to erode. In September 2012, the ice pack reached the lowest extent ever recorded, only about half the area that was covered when the *Polar Sea* reached the North Pole in 1994. The ice continues to thin as well. Now instead of multiyear ice floes several feet thick, much of the sea ice in the Arctic Ocean is first-year ice, only a foot or two thick, which readily melts away during summer.

How does the Arctic Ocean sea ice affect weather down south? Atmospheric scientists are only beginning to work this out. They know that the frozen North is a climatic "air conditioner" for North America. When summer sea ice melts away, the dark water of the Arctic Ocean absorbs more heat energy from the sun. Global heating of the planet from fossil-fuel burning also contributes extra warmth to the Arctic Ocean and atmosphere. Scientists now think that this heating of the Arctic, and the melting away of the summer ice pack, destabilizes the path of the jet stream as it crosses the United States. The erratic jet stream forms deep loops, dipping far south and allowing cold air masses normally confined to the Arctic and northern Canada to penetrate far south. In turn, the clash of cold northern air masses with warm, humid air from the Gulf of Mexico creates violent storms and precipitation, including heavy snowfalls, hail, and tornadoes. Extreme droughts, such as those experienced in Texas and the American Southwest this century, might become more common because of the climate disruptions. So the crazy weather experienced in the United States in recent years may result, in part at least, from sea ice melting in the Arctic Ocean.

Impacts on the Georgia Coast

In *Life and Death of the Salt Marsh* (1969), John and Mildred Teal expressed two main worries for these coastal systems. They saw that salt marshes in New England were disappearing because of coastal development. Ditches dug across the marsh to drain surface ponds for mosquito control, or the outright filling of marshes for pastures or buildings, were destroying these fertile habitats. At the same time, nitrogen and phosphorus nutrients were pouring into the estuaries from the fertilization of farm crops and urban lawns, and from ever-increasing volumes of treated sewage as coastal populations grew. The number of people living on Cape Cod, where the Teals worked, doubled every two decades after World War II.

There was an unintended consequence of digging drainage ditches in New England marshes to get rid of mosquito-breeding ponds on the upper marsh. The ditches were essentially new, human-made tidal creeks. Cordgrass loved that. *Spartina* plants grew green and lush along the ditches, fertilized and washed by the incoming tide. *Spartina*-eating crabs loved the new, dense growth. The squareback crab, *Sesarma reticulatum*, which nibbles tender cordgrass shoots in the low marsh, avidly moved up the drainage ditch creeks along with the creekside cordgrass. Suddenly, the squareback crabs' favorite habitat had increased, and the crabs increased in abundance along with it.

That might have been all right, but humans had caused another problem for the New England salt marshes. Since 2002, Mark Bertness, a professor at Brown University, has been working on a mysterious die-off of the creekside *Spartina* plants in these northern salt marshes. What his lab discovered was that two major predators of squareback crabs, blue crabs and striped bass, had been overfished by New Englanders for decades. The populations of these crab eaters had plummeted. As a result, the squareback population had exploded. The crabs had munched their way up the creeks and ditches, and well into the surrounding cordgrass marsh, leaving bare mudflats where once lush stands of *Spartina* grew.

Bertness and his colleagues then discovered something that might help the marsh recover. The green crab, *Carcinus maenas*, native to European coastal waters, was first noticed in the United States along the Massachusetts coast in 1817, transported most likely in the fouling communities on the bottoms of sailing ships. This three-inch-wide crab is a true invasive species: highly adaptable and subsisting on a flexible diet of small invertebrates. It quickly spread up and down the East Coast, and by 1989 had even made it to the Pacific Coast, likely on ships transiting the Panama Canal. The Bertness lab's field research showed that this voracious predator has had one positive effect in New England marshes. It scares the dickens out of the smaller squareback crabs. The green crabs have recently invaded the ditch creeks of the salt marsh in their never-ending search for food. They prey mainly on bivalves and small crustaceans along the fertile creek banks, and not so much on the cordgrass-chewing crabs. But they take over the smaller crabs' burrows for their own low-tide shelters, and they apparently bully the squareback crabs so much that they just give up and leave. Wherever the green crabs have staked a claim in the low marsh, the population of squarebacks has dived. And the cordgrass has grown back, lush and dense along the creeks and ditches.

Coastal Pollution

Southern marshes have few standing ponds of water at low tide in which mosquitoes breed. Mosquitoes in Georgia marshes take advantage of every little crevice offering a bit of moisture left by the retreating tide—fiddler crab holes and the space between *Spartina* leaves and stems—as places for their larvae to hatch and grow. So there was no reason to drain these salt marshes as the New Englanders did theirs.

But development along the southeastern coast has grown apace with development farther north. In southern coastal environments, a major problem has been plant fertilizers, from farm fields and urban lawns, coupled with sewage outflows from cities and town, which wash into the estuaries.

In 1986, Peter Verity, a researcher at the Skidaway Institute of Oceanography, located on a sea island near Savannah, Georgia, started a sampling program in the local salt marsh estuary. The plan was simple: take a bucket sample of surface water from the Skidaway Institute's dock on a tidal river every week, at both low and high tides. For each sample, Verity measured basic indicators: salinity, temperature, plant nutrients, and chlorophyll (an index of how much phytoplankton was growing in the water). No granting agency funded the work, but he thought it might make a nice record of seasonal and year-to-year changes in the estuary. Verity kept sampling, week after week, through the heat of summer and through winter storms. He kept it up for ten years. In the end, he had a decade-long record of what was going on in the estuary, week by week, season by season, year by year. This is a very valuable kind of record that granting agencies, which like to fund projects for only two to three years at a time, usually don't support.

In the meantime, while Verity was analyzing his samples, the Georgia coast was undergoing rapid development. When I first visited the Skidaway Institute on the north end of the island in the mid-1970s, the land was covered with dense oak and pine forests. Then, in the 1980s, developers discovered the island and began building fancy, high-end houses in a gated community, The Landings, at the south end. They added a small shopping center, a gas station, and a tennis court complex for the convenience of the island's homeowners. I once watched the Australian world tennis champion Evonne Goolagong play there. Through the 1980s, development on the island, as well as on the mainland across the salt marshes, increased at an ever-faster pace. House lots crept up to the north end of Skidaway, and tracts of native forest were bulldozed to create six golf courses among the newly constructed mansions. The Landings grew into one of the largest gated communities in the country.

As the population of the island and mainland increased, more sewage was treated and released into the estuaries, more lawns were lovingly tended, and the turf covering all those golf courses had to be kept lush and green with fertilizer applications. To Verity, the effects were clear. The ever-increasing load of sewage and plant nutrients inevitably found their way into his sampling bucket. Over ten years, the amounts of nitrogen and phosphorus nutrients he measured in the estuarine water nearly doubled. He also saw high spikes in nutrients after summer rainstorms lashed the island. The nutrients fueled extra phytoplankton growth. By the end of the study, summer algal blooms were twice as intense as at the start. As the phytoplankton bloomed and died, microbial decomposition used up oxygen in the water, causing more frequent anoxic "dead zones" in the estuaries. There is no doubt that the forest felling and the home and golf course building were main culprits. During the time that Verity was at the Skidaway Institute, from 1980 to 2000, the population of Skidaway Island increased by 485 percent.

Since 1990, development of the Georgia coast south of Savannah has continued, with a proliferation of large houses and golf courses. No one has yet studied whether the marshes and estuaries around Sapelo and other sea islands have been as affected as the Skidaway estuary.

An equally troubling problem for the Georgia coast is pollution by toxic chemicals from coastal industries and urban development. Long before the Georgia coast became a magnet for tourists and retirees, pulp mills were producing reams of paper and paperboard. The sulfurous "eau de pulp" emanating from the Brunswick plant, which has been reducing coastal pines to cellulosic slush since 1938, is hard to miss. Added to the toxic air and water effluents from pulp mills are wastes from chemical plants that supply chlorine bleach and other noxious chemicals for paper production. Other chemical plants cause trouble along the coast as well. In May 2011, tens of thousands of fish, as well as alligators and birds, were found dead and dying along the lower Ogeechee River. The culprit was a textile finishing plant that had been illegally discharging formaldehyde, ammonia, and hydrogen peroxide into the river for years. (In 2012, the Ogeechee Riverkeepers and other organizations were successful in getting the waste discharge permit of the textile plant revoked.)

In 1980, Congress responded to the public health disaster caused by unregulated disposal of toxic chemical wastes in the Love Canal neighborhood of Niagara Falls, New York, by enacting a law enabling the government-funded cleanup of such waste dumps. Under the Superfund law, the Environmental Protection Agency has worked with states to identify Superfund sites in need of removal of hazardous chemicals in soil and water. Coastal Georgia has four Superfund sites resulting from toxic industrial pollution. Since 1997, the Georgia Environmental Protection Division, aided by the Environmental Protection Agency, has been taking out chemical-laced sludge and monitoring toxic flows at the 84-acre Brunswick Wood Preserving site, where a former wood treatment facility used creosote, aromatic hydrocarbons such as dioxin- and furan-contaminated pentachlorophenol (PCP), and chromated copper arsenate to produce insect-resistant lumber. The 550-acre LCP Chemical site, along tidal marshes adjacent to Brunswick, is even worse. Several chemical plants operated there from the 1920s to the 1990s, leaving a chemical stew of polychlorinated biphenyls (PCBs), PCP, aromatic hydrocarbons, mercury, and other heavy metals. Two other sites are a landfill and a dredge spoil, both contaminated with organic chlorinated chemicals and lead from a plant where Hercules, Inc. manufactured pesticides. Keeping tabs on chemical cleanup and tracking the escape of pollutants into the air, land, estuaries, and waterways at these Superfund sites are major activities of the Brunswick-based Glynn Environmental Coalition (GEC). The GEC also has a Safe Seafood program to do outreach to subsistence fishers in the areas with contaminated seafood.

Toxic chemicals prevalent in one of the Brunswick Superfund sites have recently been found in local populations of bottlenose dolphins. A group of scientists from the Department of Biology and Marine Biology, at the University of North Carolina at Wilmington, were curious about the load of organic pollutants in the fatty tissue of these cetaceans along the Georgia coast. Since organic chemicals accumulate in fat, the researchers analyzed blubber samples collected from wild dolphins by shooting the animals with a retrievable dart that cut small plugs of skin and fat from their backs. Individual dolphins were recognized by dorsal fin shape and markings from photographs. The results showed high concentrations of the organic pollutants PCP and a PCB mixture, Aroclor 1268, in all the dolphins. The highest concentration of the chemicals in the blubber samples was found, not surprisingly, in dolphins swimming nearest to the LCP Chemical Superfund site in Brunswick. Since dolphins are top predators, the study pointed to the pervasiveness of Superfund pollutants in marine food webs along the Georgia coast. The researchers speculated that the burden of chemical contaminants in these long-lived marine mammals could negatively affect their reproductive and immune functions.

Pesticides used in pine forests, farm fields, and urban lawns are yet another source of pollutants that can be debilitating to coastal wildlife. One example is fipronil, a broad-spectrum insecticide used around homes and on golf courses to combat cockroaches, yellow jackets, fleas, and ever-present fire ants. Fipronil, along with other pesticides, inevitably finds its way into coastal rivers and estuaries, where it is deadly to fish and invertebrates. (This insecticide, not surprisingly, has also been linked to honeybee deaths.) Dozens of other herbicides, fungicides, and insecticides are applied along the Georgia coast. Traces of these chemicals have been detected in groundwater and estuaries; many of the toxic compounds have serious effects on the growth and reproduction of coastal wildlife.

Climate Effects

In the years since the Teals wrote about environmental dangers facing salt marshes, a new, potentially much worse threat has been recognized: climate change. When our family left Sapelo Island in 1990, atmospheric concentration of carbon dioxide, the main greenhouse gas, stood at about 350 parts per million (ppm). The world was then only dimly aware of the consequences of exceeding that level of atmospheric carbon dioxide. In May 2013, the carbon dioxide level reached the milestone of 400 ppm. Climate scientists foresee grim scenarios if that concentration continues to increase and the planet continues to warm.

Among those dire predictions are two that could severely damage the marshes and estuaries of the Georgia coast. One of these, extraordinary weather, may already be killing the marsh and some of its inhabitants.

While the New England marshes were experiencing creekside die-off in the early part of this century, southern marshes were also having episodes of "browning," the conversion of expanses of cordgrass to bare mudflats, especially in Louisiana and Georgia. Years of shortfalls of rain, along with high summer temperatures, had plagued the southeastern United States in the late 1990s and early 2000s. Researchers found that the die-off of plants in the marsh plain resulted from overstressed cordgrass that couldn't cope with the increasing saltiness of the marsh soils as the dry, hot weather of the drought evaporated water at low tide. At the same time, fewer rainstorms during the summer meant that the salt accumulating on the stressed *Spartina* leaves wasn't being washed away. The *Spartina* plants withered and died from salt stress.

But something else was contributing to the die-off. Brian Silliman, a PhD student in the Bertness lab, which had already fingered the cause of the creekside cordgrass die-off in New England marshes, went on to investigate the browning phenomenon in southeastern salt marshes. Silliman, who had studied salt marshes for years, knew a lot about the interactions of marsh plants and animals. One startling thing he had found in his research was that the abundant marsh periwinkle snail, of the genus *Littoraria*, "farmed" nutritious fungi they liked to eat. During high tide, the snails crowded on cordgrass stems to avoid predators swimming in from the creeks. They continued feeding by rasping the juicy outer layer of *Spartina* stems, making small wounds in the plant. Fungi present on the surface of the stems would grow in these rasped-out holes. The snails would repeatedly return to feed on the fungi and make the stem wounds larger. The *Spartina* plants could usually stay healthy and continue growing despite the fungal infection. But if the cordgrass plants were stressed, the fungal growth would overwhelm them and they would finally die.

Silliman and his colleagues discovered that this was the case: the fungi-farming snails played a role in the die-off of *Spartina* in at least some of the cases. In salt marshes in South Carolina, Georgia, and Louisiana, they often found huge periwinkle snail populations where the marsh was brown and bare. The researchers observed "snail fronts," hordes of hungry gastropods advancing from the barren, grazed-over marsh into adjacent healthy cordgrass stands. When snails were removed from fenced test plots, cordgrass plants started to sprout and grow back. Where the snails remained, the marsh was reduced to bare mud. Even though the drought in the southeastern United States had abated by the time of their study, Silliman and his coworkers concluded that the snails had contributed to an overwhelming fungal infection of salt-stressed marsh plants during the dry years. The snail-farmed fungus grew luxuriously on the dying cordgrass plants, yielding a huge food bonanza for the snails. After a resulting "snail boom," armies of snails swarmed over the marsh, continuing to mow down the marsh grass even after the drought was over. In some of the die-offs that Silliman studied, the grazing snail hordes extended the initial patches of barren marsh by over 200 percent. Although not all cases of marsh browning have been accompanied by snail outbreaks, the runaway consumption of cordgrass by massed snails has been observed in numerous salt marshes.

Silliman, now the Rachel Carson Associate Professor of Marine Conservation Biology at the Duke University Marine Laboratory in Beaufort, North Carolina, is continuing his research on how climate extremes can affect biotic interactions in salt marshes. These effects may lead to other unexpected harmful outcomes like the overgrazing of marsh grass by periwinkle snails.

The years of southeastern drought caused another mysterious disaster on the Georgia coast. This story is even weirder than that of the fungus-farming snails. And, curiously, it may have contributed to the explosive population increase of periwinkle snails in the salt marshes.

Blue crabs are the basis of an important fishery along the U.S. Atlantic Coast. Georgia fishermen typically land hundreds of tons of crab annually. Our family looked forward to netting blue crabs in the tidal creeks each summer. But starting in 1996, crab landings

Hordes of marsh periwinkle snails overgrazing cordgrass during a marsh die-off event. Photo provided with permission by Dr. Brian Silliman.

crashed. By 2004, the blue crab catch had nose-dived to a small fraction of the long-term average haul. What was going on? Dick Lee and Marc Frischer, colleagues of Peter Verity at the Skidaway Institute of Oceanography, decided to find out.

Many of the crabs that were caught during the decline seemed sick. Lee and Frischer knew about a particular crab disease that had caused population crashes in Europe and Alaska. The crab disease is caused by a strange blood parasite: a dinoflagellate. It turns out that this dinoflagellate, *Hematodinium*, infects and grows in the blood of marine crabs. *Hematodinium* (literally, "blood-whirler") is closely related to toxic red-tide-causing dinoflagellates and to a dinoflagellate, *Pfiesteria piscicida*, that attacks and kills estuarine fish.

Blue crabs are especially vulnerable to *Hematodinium* infection. These crustaceans don't have an immune system like ours. If

The Once and Future Coast

something strange enters a crab's bloodstream, special cells, hemocytes, engulf the intruder and eliminate it. *Hematodinium* insidiously feeds on these protective blood cells, eliminating them. The parasitic dinoflagellate also consumes the blood proteins that carry oxygen to cells throughout the crab's body. As the dinoflagellate proliferates in the crab's bloodstream, the crab slowly suffocates from lack of oxygen, becomes lethargic, and dies.

The investigators discovered that the dramatic decline in blue crab numbers after 1996 was indeed caused by severe *Hematodinium* infection. They wondered why blue crabs, which had resisted the dinoflagellate parasite in previous years, were succumbing to it now.

Hematodinium forms dinospores, resting stages that wait in coastal waters until they encounter a crab they can infect by entering the bloodstream through the crab's gills. The dinospores don't survive well when the water temperature is low, so blue crabs rarely have the disease in winter. Another environmental factor contributing to the infection of the crabs by dinospores is water salinity. Lee and Frischer found that blue crabs living in parts of the estuary that had freshwater sources, such as springs or coastal rivers, were healthy and did not appear to have the dinoflagellate parasite. But wherever estuarine waters had a salt content higher than 28 parts per thousand, crabs were virtually absent, and those that were caught were full of the blood-eating protist. Lee and Frischer speculated that the severe drought from 1997 to 2002 resulted in reduced river flows to the coast. Estuaries were flooded with saltier ocean water. *Hematodinium* flourished, and parasitic infection raged through the blue crab population as the crabs came into the Georgia estuaries in the spring to breed. Lee and Frischer speculated that since blue crabs are notorious cannibals, readily attacking and eating sickly crabs, this behavior facilitated the rapid spread of the parasitic disease.

Silliman, working out how periwinkle snails contributed to marsh browning in the same drought years, also noted the decline of the blue crabs along the southeastern coast. Marsh snails are a favorite food of blue crabs, one reason that snails flee up *Spartina* stalks at high tide as crabs swim into the marsh plain. A slump in blue crab populations would mean less predation on the snails, allowing the snails to grow freely on the dead and decaying cordgrass plants.

Other potential effects of drought on estuarine animals keep popping up. One that has been fingered is the mass die-off of oysters from dermo disease along the southeastern U.S. coast. Originally thought to be caused by a fungus, dermo is an infection by a widespread oyster parasite, *Perkinsus marinus*, named after the researcher at the Virginia Institute of Marine Science, Frank Perkins, who discovered the tiny protist in the tissues of sick oysters in Chesapeake Bay. Nonetheless, the oyster malady is still called dermo after the original, albeit incorrect, Latin name. The *Perkinsus* parasite is distantly related to the *Hematodinium* dinoflagellate that makes blue crabs ill, and like the blue crab parasite, *Perkinsus* grows in oyster blood cells, eventually taking over the whole body of the mollusk. The flesh of a heavily infected oyster turns from its normal pearly gray to a leaden blue black from the burden of protist parasites.

Investigators have long known that oysters are more susceptible to dermo during summer months, when estuarine water is warmer than 20°C (68°F) and saltier than 15 parts per thousand. A spate of droughts in the last thirty years has caused mass oyster mortality as their habitats have cooked under the hot sun, setting up perfect conditions for the dermo parasite to multiply. Mass oyster die-offs along the coasts of South Carolina and Georgia were documented in a 1986 drought. A study of the dermo disease in oysters in the Duplin River in the extremely dry years of 1999 and 2000 found that 90 percent to 100 percent of oysters collected along the length of the tidal river had *Perkinsus* reproducing in their tissues. In 2007–8, another severe drought gripped the Southeast. It was the second-driest year on record for Atlanta, Georgia; the Lake Lanier reservoir, the main water supply for that city, dropped to the lowest level ever recorded. During that drought, a group of scientists studied the impact of dermo on the productive oyster harvest of Apalachicola Bay. Decreased freshwater flow into the bay from rain-deprived rivers in Georgia caused the salt content of the water to rise to over 17 parts per thousand. The researchers found that the larger, harvestable-sized oysters suffered the greatest mortality from dermo infection as estuarine waters became saltier.

As global temperatures continue to increase, and weather patterns become more extreme, future droughts similar to the hot, dry years that plagued the Southeast at the turn of the last century seem inevitable. Then cordgrass die-offs, declines in the blue crab population, and oyster death from the dermo parasite are likely to recur. But now at least we know something about how and why these events happen.

Rising Seas

An even more worrisome consequence of a warming planet is sea-level rise. Glaciers and ice sheets at the north and south poles are wasting away; all their meltwater eventually runs into the ocean, steadily raising sea levels.

The level of the world oceans has risen and fallen many tens of feet over past geological eras. During the peak of the last glacial period, about twenty-one thousand years ago, when vast amounts of water were locked up in ice, sea levels were as much as four hundred feet lower than at present. Broad swaths of the continental shelf, now areas of shallow coastal ocean, then were dry land. Sonar soundings reveal ancient river valleys cleaving the flat expanses of offshore shelves. During the grip of the global deep freeze, sea islands and salt marshes were at the edges of the coastal shelves. Starting about nineteen thousand years ago, the world warmed and the continental ice sheets began to melt, pouring large volumes of water into the sea. The height of the invading tides rose with the sea level. Each high tide reached farther across the marsh and onto the sea islands. Eventually, the islands were completely inundated. The rise of the sea was inexorable, about three to four inches a decade, three feet a century. During some periods, more abrupt rises occurred: up to eight feet a century, equivalent to about ten inches every decade. The marshes and sea islands retreated inland or disappeared altogether for a time. After sea levels stabilized as the last continental ice sheets melted, the present location of marshes and islands was finally established. Over the last five thousand years, the Georgia coastline has looked pretty much as it does today.

Increasing global temperatures, due to the accumulation of greenhouse gases in the atmosphere, are set to disrupt the pleasant balance of clement weather and constant sea levels under which human civilizations have developed and thrived. According to the *Fifth Assessment Report of the International Panel on Climate Change*, the projected increase in sea levels could be as much as three feet by 2100, a pace not seen since the end of the last glacial period. Once the global climate reaches a tipping point beyond which the melting of the polar ice caps is inevitable, the continuing rise in sea height will be unstoppable.

In their book *The Rising Sea*, Orrin Pilkey and Rob Young discuss the increase in sea levels expected if there were a complete disintegration of these ice sheets. A total loss of the ice sheets covering Greenland would result in a rise of 24 feet. Melting of the two great ice sheets on Antarctica has the potential to increase sea levels by a staggering 187 feet. The rise caused by all polar ice sheets melting would be over 200 feet. Pilkey and Young caution that while scientists can't predict how quickly the ice sheets might release so much water into the ocean, we can't count on a uniform rate of sea-level increase. Ice sheet disintegration can be nonlinear, speeding up during some periods, perhaps to a foot per decade.

The marshes and sea islands of the Georgia coast are already vulnerable to weather events such as hurricanes as destructive as the disastrous storm of 1898. Rising sea levels will add to their vulnerability. The highest land on Georgia sea islands is only about twenty-five feet above the present mean sea level. A rise in ocean height of just three feet would inundate the island beaches and surrounding marshes. Further increases would eat away at the bordering trees and shrubs, killing them and washing away the sandy soil. Coastal freshwater ponds and aquifers will be first tainted and then swamped by the salt in the invading sea. The sea islands will gradually become uninhabitable for people, and then for the land mammals, reptiles, and birds that call them home.

Southeastern sea islands and marshes may retreat inland, as happened in past eras, although coastal habitats will remain unstable as long as sea height continues to rise at a rate unprecedented in recent geological history. There are varying estimates of the maximum annual rate of sea-level rise beyond which salt marshes would drown. Along the coast of Massachusetts, the tidal range is low and little sediment flows into the fringing salt marshes. There, a sea-level rise greater than five millimeters, about a fifth of an inch, per year would flood the salt marshes faster than they could retreat. For the southeastern coast of the United States, where the tidal range is much greater and loads of silt and clay are deposited on the marshes by coastal rivers, the marshes would be more resilient. Models suggest that Georgia salt marshes could likely keep up with an annual sea level increase averaging ten millimeters (one centimeter), or about two-fifths of an inch. A faster rate of sea-level rise would drown and eventually degrade and wash away the marshes, leaving bare mudflats where once verdant cordgrass meadows nurtured a thriving food web. At the same time, beaches and dunes would steadily erode on the ocean side of the islands, succumbing to higher and higher tides.

A rise in sea height of three feet, or about one meter, during this century is equivalent to an average rate of just over ten millimeters (one centimeter) a year. Current sea-level rise is about three millimeters a year; the rate is bound to increase as the Greenland and West Antarctic ice sheets melt away. Pilkey and Young suggest that a three-foot rise in sea level is probably a minimum estimate; the rise could be as much as six feet by the end of this century. They also worry that in some regions, coastal development has robbed marshes and beaches of space to migrate into. This is a grim scenario for these vibrant, bountiful ecosystems. The loss of coastal marshes is a problem not limited to the Georgia shore; in their 2009 review "Impacts of Global Climate Change and Sea-Level Rise on Tidal Wetlands," J. Court Stevenson and Michael S. Kearney concluded that if global warming does result in a sea-level rise of more than three feet by 2100, "as much as 90 percent of the tidal marshes worldwide could be in jeopardy."

Understanding the future dangers to the Georgia coast should only increase our appreciation of the myriad microbes, plants, and animals that live here. We can and should take on the responsibility of continuing our protection of these splendid wild habitats so that future generations can enjoy them as we do now. Locally, we can ensure that coastal development does not degrade the sea islands and marshes. An appendix at the end of the book lists conservation groups working to protect, and educate the public about, the natural environments of the Georgia coast. These groups welcome donations and volunteers. Nationally and internationally, we can support efforts to limit the increase of greenhouse gases and thus halt the progress of climate change leading to extreme droughts, storms, and sea-level rise. The little animals in the marsh muds, and the mummichogs in the tidal creeks, in the end are depending on us.

ACKNOWLEDGMENTS

A great deal of the information in this book comes from research papers in the Collected Reprints of the University of Georgia Marine Institute. These would not be available without the unflagging effort of the UGMI librarians: Lorene Townsend Howard, who was in charge of the Marine Institute's collection when Barry and I lived on Sapelo Island, and who gave us a complete set of the UGMI Collected Reprints when our family left in 1990; and the current UGMI librarian, Laura Cammon, who has put a list of all UGMI reprints on the UGMI website, including PDF copies of many of these papers. I am indebted to Lloyd Dunn, who did research on the salt marsh plant production and who kindly took me up in a small plane to photograph the island and surrounding marshes from the air. The draft manuscript received critical fact-checking by my husband, Barry Sherr, and my brother, David Brown. I especially thank Denise Seliskar, Jack Gallagher, and Robert Christian for their help with questions about coastal plant distributions and species names. Daniel Parshley, project manager of the Glynn Environmental Coalition, helped with information on Georgia's coastal Superfund sites. Brian Silliman kindly provided information about, and photographs of, periwinkle snails overgrazing *Spartina* marsh. Richard Heard provided additional useful information on marsh invertebrates. Clay Montague and another reviewer greatly improved the book by providing exceptionally detailed and thoughtful reviews of a draft manuscript. The guidance of the expert manuscript shepherds at the University of Georgia Press—Patrick Allen, Elizabeth Crowley, and Jon Davies, among others—as well as the efficient indexing service provided by Jennifer Spanier, greatly facilitated the publication process.

APPENDIX 1
Where to Go to Enjoy Georgia Coastal Wildlife

Colonial Coast Birding Trail

This trail, which has detailed information on its website, includes a number of excellent sites along the Georgia coast where people can explore the habitats and animals featured in this book. Many of these sites highlight aspects of the long and fascinating history of this region.

Fort Pulaski National Monument

East of Savannah on Highway 80, just before Tybee Island, is a 5,600-acre preserve surrounding a historic fort at the mouth of the Savannah River, managed by the National Park Service. Although the focus of the monument is the history of the fort, the site of a Civil War battle, the preserve's islands and surrounding salt marshes are prime sites for spotting coastal birds and other wildlife. Painted buntings are common in spring and summer.

Tybee Island North Beach

The beach of the northernmost island on the Georgia coast is easily reached from Savannah via Highway 80 and is a popular tourist destination, with beachside hotels and restaurants. The North Beach Birding Trail, near the Tybee lighthouse overlooking the Savannah River, is known for shorebirds, especially black skimmers. Bottlenose dolphins can be spotted playing just offshore. In fall and winter, migrating flocks of ducks and sandpipers throng the river and beaches. Gannets and Caspian terns are common in winter. The North Beach is also a great place for beachcombers looking for shells and shark teeth washed up onshore.

Little Tybee Island

This pristine nature reserve south of Tybee Island is operated by the State of Georgia. The reserve offers expanses of salt marshes, woodlands, and undisturbed beaches to explore. While there are no accommodations on the island, camping is allowed. Little Tybee can be reached only by boat; local charter services offer day tours or overnight camping expeditions to the island. Birders can look for white ibis, roseate spoonbills, ospreys, and bald eagles.

Skidaway Island State Park

This 588-acre park on the west side of Skidaway Island, reached by the Diamond Causeway, abuts Georgia's Intracoastal Waterway. Maritime forest and salt marsh habitats can be explored along two hiking trails. A boardwalk over the salt marsh ends at an observation tower. Marsh songbirds and wading birds can be seen all year; waterfowl are most abundant in winter. The area is noted for bald eagles and migrating warblers. The park has picnicking, tent and RV camping, an interpretive center, and ranger programs.

Harris Neck National Wildlife Refuge (NWR)

This large, 2,842-acre reserve in McIntosh County was the site of a World War II army airfield. The refuge is easily reached from Interstate 95. Take exit 67 off I-95 and travel south one mile on U.S. 17, then turn east on Harris Neck Road for seven miles to the main entrance. The refuge's salt marshes, forested wetlands, and open fields can be accessed by four miles of paved roads and by several hiking trails. Egrets, herons, and wood storks nest in the freshwater ponds. Note that pets are not allowed in the NWR.

Sapelo Island National Estuarine Research Reserve

The Georgia Department of Natural Resources (DNR) manages the wild habitats of Sapelo Island. A visitor center at Meridian Dock, eight miles north of Darien off Highway 99, provides information on the island, and wildlife can be spotted in the marshes and creek at the dock. The DNR runs half-day tours to Sapelo Island by ferry on Wednesday and Saturday all year, and also on Fridays from June through August. Tours on other days of the week can be arranged with the Sapelo Island guides R. J. Grovner (a former playmate of our sons) and Maurice Bailey. Overnight stays on the island are available at the Reynolds Mansion (Big House) and also by renting private guest homes. A stay with Cornelia and Julius Bailey at their six-room lodge, The Wallow, in the Hog Hammock community is a special treat for island visitors. Their son, Maurice Bailey, also operates a campground in Hog Hammock.

Tolomato Island

This nature preserve community of single-family homes on the coast, located just off Highway 99 north of Darien, features a freshwater pond and walking trails popular with birders. Ibis, heron, and wood storks nest in trees around the pond. Roseate spoonbills and black-bellied whistling ducks have also been spotted. Tabby ruins of an old sugar mill and rum distillery are near the entrance to the island community. The causeway to the island passes salt marshes and tidal creeks.

Hofwyl-Broadfield Plantation State Historic Site

This state park on the lower Altamaha River features buildings of an antebellum rice plantation, along with historical explanations of the site. A nature trail among the oak trees, cypress swamps, and river marshes offers opportunities to spot birds and other wildlife of freshwater tidal habitats. The site is between Brunswick and Darien on U.S. Highway 17, one mile east of I-95 from exit 42.

Jekyll Island

This developed island is linked to the mainland by the Jekyll Island Causeway, reached via exit 29 on I-95, then Highway 17 east for 5.4 miles. About four miles down the causeway is the Visitor's Center, from which salt marshes and mudflats can be scanned for coastal wildlife. Two miles farther east is the entrance to the island, which requires a visitor's parking fee (good for twenty-four hours). A worthwhile stop is the Tidelands Nature Center on Riverview Drive, a 4-H facility operated by the University of Georgia. There is a small admission fee, and it is open weekdays 9 a.m.–4 p.m., weekends 10 a.m.–2 p.m. Several exhibit areas showcase saltwater animals, seashells, coastal reptiles and birds, and an outdoor shark exhibit and nature trail. Live animals on display include young loggerhead turtles, baby alligators, snakes, turtles, crabs, and fish. Guides lead public nature walks and tours of the island. The Georgia Sea Turtle Center, located at 214 Stable Road on Jekyll Island, focuses on sea turtle rehabilitation, research, and education programs. The center is open to the public for several hours on most days. South Beach, at the southern tip of the island on Beachview Drive, is a good place for beachcombing and spotting shorebirds. From the parking lot next to the soccer complex, walk across back dunes and interdune meadows on the Glory Boardwalk, built during filming of the Civil War movie *Glory* for a climactic battle scene. Farther north on Beachview Drive is Great Dunes Park, which also has a boardwalk to the beach. At the northern end of the island is Driftwood Beach, which has vistas of salt marshes, ghostly dead trees on the beach, and St. Simons Sound extending from the Clam Creek Picnic Area. Several other beach-access points are at designated parking areas along North Beachview Drive. Visit the Jekyll Island Authority website for visitor information: http://www.jekyllisland.com/jekyll-island-authority. Taylor Schoettle's book *A Guide to a Georgia Barrier Island* offers in-depth information about visiting natural habitats on Jekyll.

Cumberland Island National Seashore

The largest sea island on the Georgia coast has 9,800 acres of undeveloped wilderness and pristine ocean beaches. Wildlife abounds, and sea turtles nest on the beach. As with Little Tybee and Sapelo, the island can be reached only by ferry. Check out the National Park Service's informative website for information about island activities and the ferry schedule (www.nps.gov/cuis/index.htm). To get to the ferry dock, take Exit 3 from I-95 and go east on Highway 40 to St. Mary's, Georgia. Turn right on St Mary's Street; a blue building on the left houses the NPS visitor's center at Dungeness Dock. In 2013, fare for a round trip to the island was $20 per adult, $14 for children twelve and under. There is also a small per-person fee for admission to the park, and a per-night camping fee. Ferry and camping reservations are recommended. Note that pets and bicycles are not allowed on the ferry. Once on the island, visitors can take trails from the ferry dock to salt marshes, a maritime forest, ocean beaches, and historic ruins, with restrooms and drinking water available. Tent camping on the island for up to seven nights is permitted.

APPENDIX 2

Conservation Organizations Working to Protect the Georgia Coast

These nonprofit education and advocacy groups have informative websites describing their goals and activities; they welcome donations, new members, and volunteers.

Center for a Sustainable Coast

Based on St. Simons Island, a group of professionals and citizens who advocate for the protection of Georgia coastal resources through education, advice, and legal action.

Glynn Environmental Coalition

An environmental organization in Brunswick focused on improving the health of coastal ecosystems and residents by raising awareness of industrial air and water pollution. The coalition monitors the cleanup of toxicants at Superfund sites in Glynn County, as well as contamination of well water and seafood by toxic chemicals. Its Safe Seafood program does outreach to subsistence fishers in coastal areas known to have contaminated seafood. The group is especially concerned about the effects of toxic chemicals in the environment on women of childbearing age and young children. It provides testing of hair samples to check for the presence of hazardous chemicals in the body through the University of Georgia Marine Extension Service.

Ogeechee Audubon Society

The Savannah chapter of the National Audubon Society educates all age groups about birds and the environment. Members receive a bimonthly newsletter, the *Marshlander*, and the society offers free birding trips and nature lectures that are open to the public.

Coastal Group, Part of the Georgia Chapter of the Sierra Club

Serving residents of Savannah, the chapter engages in advocacy and legal action to protect coastal wetlands.

Southern Environmental Law Center—Georgia Coastal Initiative

The Southern Environmental Law Center (SELC) works in six southeastern states, using the law to protect air and water resources. The SELC's Georgia Coast Initiative employs conservation advocacy and legal strategies to protect coastal estuaries and marshes.

Waterkeeper Alliance Affiliate Groups

In 1983, the first Riverkeeper alliance was started by commercial and sport fishermen to monitor the health of the Hudson River in New York. In 1999, the Waterkeeper Alliance was founded to support regional programs to combat the pollution of waterways across the United States. Four Riverkeeper organizations in Georgia monitor and protect freshwater streams flowing into Georgia estuaries:

Savannah Riverkeeper in Augusta
Ogeechee Riverkeeper in Statesboro
Altamaha Riverkeeper in Darien
Satilla Riverkeeper in Woodbine

Sapelo Foundation

In addition to these groups, the Sapelo Foundation, a charitable organization founded by R. J. Reynolds, who owned Sapelo Island, gives grants to projects that promote environmental protection and social justice within the state of Georgia.

APPENDIX 3
Methods for Collecting and Inspecting Coastal Biota

Benthic Algae

During low tide, benthic diatoms living in marsh mudflats migrate to the surface to carry out photosynthesis. One can take advantage of this behavior to study these microalgae. All one needs is two sheets, about six inches square, of a fine-meshed material that can stand up to wetting. Lens paper or a white close-weave cloth or netting will do. Also take a pint container of seawater, either collected from the tidal creek or artificial seawater made by dissolving a teaspoon of salt in a pint of tap water. At low tide, find a creek bank or mudflat that has a golden brown color from the diatoms basking in the sun. You may be able to reach the mudflat from a boardwalk over the marsh. If you need to step out on the mudflat, be sure it is firm enough to hold your weight.

Wet a piece of the material with the salt water and gently lay it onto the surface of the mud. Wet a second piece and position it over the top of the first layer. Now for the hard part: patiently wait for about half an hour as the diatoms, finding themselves shaded by the fabric, move up into the top layer. The longer you wait, the more diatoms you will collect. But keep an eye on the incoming tide! Finally, carefully peel off the top layer of material, which should now have a brown color, and put it into the container of seawater. Swish the material around to wash off the diatoms. A drop of this water will contain thousands of pennate diatoms, rapidly gliding about. The shapes of the diatoms' glass shells can be viewed close up with a hand lens or microscope.

Other benthic microbes can be collected by finding dense colonies of cells and scraping samples into small containers. Patches of blue-green cyanobacteria in the marsh and blooms of green euglenae on the beach can be sampled in this way. Add a bit of saltwater to the scraping and gently shake to separate the lighter algal cells from the heavier mud or sand. Then transfer a small amount of water to a dish to view with a hand lens or under a microscope.

Zooplankton

A gallon of water taken from the estuary will usually have only a couple of zooplankton swimming about. To appreciate the diversity of minute creatures in the plankton, you must concentrate their numbers. The tried-and-true method for collecting zooplankton is to screen large quantities of seawater through a fine-mesh plankton net. Plankton nets can be purchased from biological supply companies such as Carolina Biological. But commercial nets are pretty expensive, and you can easily make your own net from everyday household items.

You will need panty hose (used will do if at least one of the hose legs is in good shape), scissors, a wire coat hanger (or a wooden needlepoint hoop), a clean pint glass jar (such as a Mason jar), duct tape, needle and thread, sturdy string, and a one- or two-ounce lead fishing weight. First, tie off one of the panty hose legs (the one with the most runs if you have a used pair) close to the top with string, and cut off the excess. Bend the coat hanger into an approximately circular shape. Stretch the top of the panty hose around the wire coat hanger. Secure the panty hose to the coat hanger by stitching loosely around with thread, then reinforce it by putting duct tape over the top of the hose around the wire. (If using a needlepoint hoop, pull the top of the pantyhose through the open hoop and then secure the hose by clamping the hoop down on the hose. You should still line the top of the hose with duct tape for extra strength.) Using the narrow point of the scissors, punch three holes through the duct tape just under the coat hanger wire (or hoop), at equidistant intervals. Tie one two-foot length of string through each of the three holes. Tie the three lengths of string securely together at the other end. Make sure all these knots are good and tight. A lead weight can be tied to the knot at the top of the net to help the net sink down into the water. Around the terminal knot, tie on a longer length of string or cord, ten to twenty feet long. This will be used to lower your net into the water. Finally, cut off the foot of the panty hose leg and fasten the clean pint jar to the end of the net (the cod end) with duct tape. Voila! You have made your very own plankton net.

Plankton samples are usually collected by towing a plankton net at low speed behind a boat. Along the Georgia coast, tidal currents are strong enough that a plankton sample can be taken by lowering the net from a dock. You can fasten the net to the dock and leave it while the tide is running. Just be sure that the water is deep enough that the net doesn't hit bottom and that the net won't hang up on the dock or on boats. The entire net must sink below the surface. A good technique is to first fill the jar on the net with seawater and then lower the net straight down into the water before it stretches out into the current. This prevents air from being trapped in the net.

Since zooplankton tend to be more active at night, plankton tows made near or just after sunset will collect more creatures than tows taken during the day. Leave the net in the water for a half hour or more. At the end of the tow, haul the net in so the jar remains upright. Push the sample jar up through the middle of the panty hose and gently pour the contents into another clean glass jar. Hold the jar up to the light. You should see lots of small, darting "bugs" zipping about. These are copepods, miniature crustaceans with long antennae. There may also be several small pulsing jellyfish, hydromedusae, or a glittering ctenophore. For a good look at these interesting animals, study your catch under a magnifying glass, hand lens, or low-power microscope. The zooplankton can be slowed down by adding some club soda to the sample. The carbon dioxide in the soda will make the little animals drowsy.

Wash off the net and jar with fresh water while the net is still wet, and let dry to use for another zooplankton sample. (The hook on the coat hanger is useful for hanging the net up to dry.)

Small Fish and Shrimp

Free-swimming animals in the estuary, such as small fish and shrimp, can be collected by seining in shallow water. The use of seines less than twelve feet long, and with a mesh size of less than one square inch, are allowed in Georgia saltwater environments. A small seine can be bought. Or one can easily make a simple seine from two broom handles, rope, a length of netting, Styrofoam balls, and lead weights. The netting should be of strong cord or line, with quarter-inch to half-inch openings (measure the openings by stretching the net fabric), two to three feet wide, and from six to ten feet long. Run lengths of quarter-inch rope through the top and bottom rows of mesh in the netting. To the top length of rope, fasten two-inch-diameter Styrofoam balls every few inches (this step is not really necessary if the seine is going to be used in water shallower than the seine is high). To the bottom rope, attach lead weights of the size used in cast nets every few inches with fishing line. The weights are needed to keep the bottom of the seine on the sediment surface. Wrap a short section of each side of the net around a broom handle, and tie the ropes at the top and bottom of the net to the broom handles.

Two people are needed to operate a seine, one at each end. Both handlers can wade out a distance from shore, then slowly pull the net to the beach, maintaining the net in a U-shape and carefully keeping the bottom of the net next to the bottom. Or one person can stand on the shore and the other wade out in a curving arc from one side to the other. After gently inspecting the small fish you collect, release them back into the sea.

Benthic Animals

Tiny benthic animals, the meiofauna, can be collected by using a plankton net or a small-mesh kitchen sieve. Put a scoopful of surface marsh mud or wet beach sand into a plastic bucket about two-thirds full of seawater. Stir the mud or sand in the water. Small animals will be dislodged from the sediment. Let the mud or sand settle down, then pour the overlying water (with the animals) through the net or sieve. The animals will be concentrated in the glass jar at the bottom of the net, or on the surface of the sieve. Larger benthic critters such as clams and worms need to be dug up out of the mud or sand, but since deep excavation can destroy the burrow structure of these animals, it is not recommended.

BIBLIOGRAPHY

SUGGESTED BOOKS FOR FURTHER READING

Bailey, Cornelia W., and Christena W. Bledsoe. *God, Dr. Buzzard, and the Bolito Man: A Saltwater Geechee Talks about Life on Sapelo Island.* New York: Doubleday, 2000.

Bryant, David, and George Davidson. *Georgia's Amazing Coast.* Athens: University of Georgia Press, 2003.

Clayton, Tonya D., Lewis A. Taylor, William J. Cleary, Paul E. Hosier, Peter H. F. Graber, and Orrin H. Pilkey Sr. *Living with the Georgia Shore.* Durham, N.C.: Duke University Press, 1992.

Dahlberg, Michael. *Guide to Coastal Fishes of Georgia and Nearby States.* 1975. Reprint, Athens: University of Georgia Press, 2008.

Gibson, Count D. *Sea Islands of Georgia: Their Geologic History.* Athens: University of Georgia Press, 2010.

Johnson, A. Sydney, Hilburn O. Hillestad, Sheryl F. Shanholtzer, and G. Frederick Shanholtzer. *An Ecological Survey of the Coastal Region of Georgia.* National Park Service Scientific Monograph 3, 1974. http://www.npshistory.com/series/science/3/index.htm.

Kleppel, Gary S., M. Richard DeVoe, and Mac V. Rawson, eds. *Changing Land Use Patterns in the Coastal Zone: Managing Environmental Quality in Rapidly Developing Regions.* New York: Springer, 2006.

Lenz, Richard J. *Longstreet Highroad Guide to Georgia Coast and Okefenokee.* N.p.: John F. Blair, 2002. http://www.sherpaguides.com/georgia/coast/.

Martin, Anthony J. *Life Traces of the Georgia Coast: Revealing the Unseen Lives of Plants and Animals.* Bloomington: Indiana University Press, 2013.

McKee, Gwendolyn, ed. *A Guide to the Georgia Coast: The Georgia Conservancy.* Atlanta: Longstreet, 1993.

Pomeroy, Lawrence R., and Richard Wiegert, eds. *The Ecology of a Salt Marsh.* New York: Springer-Verlag, 1981.

Ruppert, Edward, and Richard S. Fox. *Seashore Animals of the Southeast: A Guide to Common Shallow-Water Invertebrates of the Southeastern Atlantic Coast.* Columbia: University of South Carolina Press, 1988.

Schoettle, Taylor. *A Guide to a Georgia Barrier Island: Featuring Jekyll Island with St. Simons and Sapelo Islands.* Darien, Ga.: Watermarks, 1996.

Seabrook, Charles. *The World of the Salt Marsh: Appreciating and Protecting the Tidal Marshes of the Southeastern Atlantic Coast.* Athens: University of Georgia Press, 2012.

Silliman, Brian R., Mark D. Bertness, and Edwin D. Grosholz, eds. *Human Impacts on Salt Marshes: A Global Perspective.* Berkeley and Los Angeles: University of California Press, 2009.

Sullivan, Buddy. *Early Days on the Georgia Tidewater: The Story of McIntosh County and Sapelo.* Darien, Ga.: McIntosh County Board of Commissioners, 1997.

Teal, John, and Mildred Teal. *Life and Death of the Salt Marsh.* New York: Ballantine, 1969.

Teal, Mildred, and John Teal. *Portrait of an Island.* Athens: University of Georgia Press, 1967. Reprint, 1997.

Wilson, Jim. *Common Birds of Coastal Georgia.* Athens: University of Georgia Press, 2011.

CHAPTER SOURCES

Preface. Island Bound

Kneib, Ronald T. "The University of Georgia Marine Institute." *Georgia Journal of Science* 54 (1996): 81–89.

Odum, Eugene P. "The Strategy of Ecosystem Development." *Science* 164, no. 3877 (1969): 262–70.

Odum, Eugene P., and Howard T. Odum. *Fundamentals of Ecology.* Philadelphia: Saunders, 1953.

Schnakenberg, Heidi. *Kid Carolina: R. J. Reynolds Jr., a Tobacco Fortune, and the Mysterious Death of a Southern Icon.* New York: Center Street, 2010.

Sullivan, Buddy. "The Historic Buildings of Sapelo: A 200-Year Architectural Legacy." Occasional Papers of the Sapelo Island National Estaurine Research Reserve, Vol. 2, 2010.

Chapter 1. Marine Habitats of the Georgia Coast

Chalmers, Alice G. *The Ecology of the Sapelo Island National Estuarine Research Reserve.* N.p.: NOAA Office of Coastal Resource Management and Georgia Department of Natural Resources, 1997. http://www.sapelonerr.org/wp-content/uploads/2013/08/Ecology-of-the-Sapelo-Island-National-Estuarine-Research-Reserve.pdf.

Frey, Robert W., and Paul Basan. "Coastal Salt Marshes." In *Coastal*

Sedimentary Environments, 2nd ed., edited by Richard A. Davis Jr., 225–301. New York: Springer-Verlag, 1985.

Gibson, Count D. *Sea Islands of Georgia: Their Geologic History*. Athens: University of Georgia Press, 2010.

Haines, Bruce L., and E. Lloyd Dunn. "Coastal Marshes." In *Physiological Ecology of North American Plant Communities*, edited by Brian F. Chabot and Harold A. Mooney, 323–47. London: Chapman and Hall, 1985.

Hoyt, John H. "Barrier Island Formation." *Geological Society of America Bulletin* 78 (1967): 1125–36.

Hoyt, John H., and Robert J. Weimer. "Comparison of Modern and Ancient Beaches, Central Georgia Coast." *Bulletin of the American Association of Petrology and Geology* 47 (1963): 529–31.

Martin, Anthony J. *Life Traces of the Georgia Coast: Revealing the Unseen Lives of Plants and Animals*. Bloomington: Indiana University Press, 2013.

Pilkey, Orrin H., and Dennis M. Richter. "Beach Profiles of a Georgia Barrier Island." *Southeastern Geology* 6 (1964): 11–19.

Pomeroy, Lawrence R. "The Ocean's Food Web: A Changing Paradigm." *BioScience* 24 (1974): 499–504.

Pomeroy, Lawrence R., and Richard G. Wiegert, eds. *The Ecology of a Salt Marsh*. Ecological Studies 38. New York: Springer-Verlag, 1981.

Teal, Mildred, and John Teal. *Portrait of an Island*. Athens: University of Georgia Press, 1967. Reprint, 1997.

Chapter 2. What You Don't See: Microscopic Life

Bergh, Oivind, Knut Y. Borsheim, Gunnar Bratbak, and Mikal Heldal. "High Abundance of Viruses Found in Aquatic Environments." *Nature* 340, no. 6233 (1989): 467–68.

Christian, Robert R., and William J. Wiebe. "Anaerobic Microbial Community Metabolism in *Spartina alterniflora* Soils." *Limnology and Oceanography* 23 (1978): 328–36.

Goldstein, Susan T., and Robert W. Frey. "Salt Marsh Foraminifera, Sapelo Island, Georgia." *Senckenbergiana Maritima* 18 (1986): 97–121.

Howarth, Robert W., and Roxanne Marino. "Sulfate Reduction in Salt Marshes, with Some Comparisons to Sulfate Reduction in Microbial Mats." In *Microbial Mats: Stromatolites*, edited by Yehuda Cohen, Richard W. Castenholz, and H. Orin Halvorson, 245–63. New York: Liss, 1984.

Howarth, Robert W., and Susan Merkel. "Pyrite Formation and the Measurement of Sulfate Reduction in Salt Marsh Sediments." *Limnology and Oceanography* 29 (1984): 598–608.

Johannes, Robert E. "Influence of Marine Protozoa on Nutrient Regeneration." *Limnology and Oceanography* 10 (1965): 434–42.

———. "Phosphorus Excretion and Body Size in Marine Animals: Microzooplankton and Nutrient Regeneration." *Science* 146, no. 3646 (1964): 923–24.

Odum, Eugene P., and Armando A. de la Cruz. "Detritus as a Major Component of Ecosystems." *American Institute of Biological Sciences Bulletin* 13 (1963): 39–40.

Pomeroy, Lawrence R., and Robert E. Johannes. "Total Plankton Respiration." *Deep-Sea Research* 13 (1966): 971–73.

Richardson, Joseph P. "Floristic and Seasonal Characteristics of Inshore Georgia Macroalgae." *Bulletin of Marine Science* 40, no. 2 (1987): 210–19.

Sherr, Evelyn B., and Barry F. Sherr. "Understanding Roles of Microbes in Marine Pelagic Food Webs: A Brief History." Chapter 2 of *Microbial Ecology of the Oceans*, 2nd ed., edited by David L. Kirchman, 27–44. New York: Wiley-Liss, 2008.

Teal, John M. "Energy Flow in the Salt Marsh Ecosystem of Georgia." *Ecology* 43 (1962): 614–24.

Chapter 3. Marsh Grass, Live Oaks, Sea Oats

Bradley, Paul M., and E. Lloyd Dunn. "Effects of Sulfide on the Growth of Three Salt Marsh Halophytes of the Southeastern United States." *American Journal of Botany* 76, no. 12 (1989): 1707–13.

Duncan, Wilbur H. *The Vascular Vegetation of Sapelo Island, Georgia*. Athens: University of Georgia Botany Department and Georgia Department of Natural Resources, 1982.

Gallagher, John L. "Zonation of Wetland Vegetation." In *Coastal Ecosystem Management*, edited by John R. Clark, 752–58. New York: Wiley-Interscience, 1977.

Johnson, A. Sydney, Hilburn O. Hillestad, Sheryl F. Shanholtzer, and G. Frederick Shanholtzer. *An Ecological Survey of the Coastal Region of Georgia*. National Park Service Scientific Monograph 3, 1974. http://www.nps.gov/history/history/online_books/science/3/index.htm.

Kellogg, Elizabeth A. "Evolutionary History of the Grasses." *Plant Physiology* 125 (2001): 1198–205.

Teal, Mildred, and John Teal. *Portrait of an Island*. Athens: University of Georgia Press, 1967. Reprint, 1997.

Chapter 4. Creatures of the Black Goo

Coull, Bruce C., and Bettye W. Dudley. "Dynamics of Meiobenthic Copepod Populations: A Long-Term Study (1973–1983)." *Marine Ecology Progress Series* 24 (1985): 219–29.

Darling, John A., Adam R. Reitzel, Patrick M. Burton, Maureen E. Mazza, Joseph F. Ryan, James C. Sullivan, and John R. Finnerty. "Rising Starlet: The Starlet Sea Anemone, *Nematostella vectensis*." *BioEssays* 27 (2005): 211–21.

Healy, Brenda, and Keith Walters. "Oligochaeta in *Spartina* Stems: The Microdistribution of Enchytraeidae and Tubificidae in a Salt Marsh, Sapelo Island, USA." *Hydrobiologia* 278 (1994): 111–23.

Hicks, G. R. F., and Bruce C. Coull. "The Ecology of Marine Meiobenthic Harpacticoid Copepods." *Oceanography and Marine Biology Annual Review* 21 (1983): 67–175.

Kneib, Ronald T. "Population Dynamics of the Tanaid *Hargeria rapax* (Crustacea: Peracarida) in a Tidal Marsh." *Marine Biology* 113 (1992): 437–45.

Rieper, M. "Feeding Preferences of Marine Harpacticoid Copepods for Various Species of Bacteria." *Marine Ecology Progress Series* 7 (1982): 303–7.

Walters, Keith, Erin Jones, and Lisa Etherington. "Experimental Studies of Predation on Metazoans Inhabiting *Spartina alterniflora* Stems." *Journal of Experimental Marine Biology and Ecology* 195, no. 2 (1996): 251–65.

Chapter 5. Mud Dwellers of Marshes and Creeks

Bertness, Mark, and Edwin Grosholz. "Population Dynamics of the Ribbed Mussel, *Geukensia demissa*: The Costs and Benefits of an Aggregated Distribution." *Oecologia* 67 (1985): 192–204.

Frankenberg, Dirk, and William D. Burbanck. "A Comparison of the Physiology and Ecology of the Estuarine Isopod *Cyathura polita* in Massachusetts and Georgia." *Biological Bulletin* 125 (1963): 81–95.

Kneib, Ronald T., and Cynthia A. Weeks. "Intertidal Distribution and Feeding Habits of the Mud Crab, *Eurytium limosum*." *Estuaries* 13, no. 4 (1990): 462–68.

Kraeuter, John N. "Biodeposition by Salt-Marsh Invertebrates." *Marine Biology* 35 (1976): 215–23.

Lin, Junda. "Influence of Location in a Salt Marsh on Survivorship of Ribbed Mussels." *Marine Ecology Progress Series* 56 (1989): 105–10.

MacKenzie, Clyde L., Jr. "History of Oystering in the United States and Canada, Featuring the Eight Greatest Oyster Estuaries." *Marine Fisheries Review* 58, no. 4 (1996): 1–78.

Martin, Anthony J. *Life Traces of the Georgia Coast: Revealing the Unseen Lives of Plants and Animals.* Bloomington: Indiana University Press, 2013.

Montague, Clay L. "A Natural History of Temperate Western Atlantic Fiddler Crabs (Genus *Uca*) with Reference to Their Impact on the Salt Marsh." *Contributions in Marine Science* 23 (1980): 25–55.

———. "The Influence of Fiddler Crab Burrows and Burrowing on Metabolic Processes in Salt Marsh Sediments." In *Estuarine Comparisons*, edited by Victor S. Kennedy, 283–301. New York: Academic Press, 1982.

Pace, Michael L., Stephen Shimmel, and W. Marshall Darley. "The Effect of Grazing by a Gastropod, *Nassarius obsoletus*, on the Benthic Microbial Community of a Salt Marsh Mudflat." *Estuarine and Coastal Marine Science* 9, no. 2 (1979): 121–34.

Ruppert, Edward E., and Richard S. Fox. *Seashore Animals of the Southeast: A Guide to Common Shallow-Water Invertebrates of the Southeastern Atlantic Coast.* Columbia: University of South Carolina Press, 1988.

Smith, Jennifer M., and Robert W. Frey. "Biodeposition by the Ribbed Mussel *Geukensia demissa* in a Salt Marsh, Sapelo Island, Georgia." *Journal of Sedimentary Research* 55, no. 6 (1985): 817–28.

Sullivan, Buddy. "The Historic Buildings of Sapelo: A 200-Year Architectural Legacy." Occasional Papers of the Sapelo Island National Estuarine Research Reserve, vol. 2, 2010.

Teal, John M. "Distribution of Fiddler Crabs in Georgia Salt Marshes." *Ecology* 39 (1958): 186–93.

Thoresen, Merrilee, Merryl Alber, and Randal L. Walker. "Trends in Recruitment and *Perkinsus marinus* Parasitism in the Eastern Oyster, *Crassostrea virginica*, within the Sapelo Island National Estuarine Research Reserve (SINERR)." Marine Technical Report no. 05-1. Athens: University of Georgia School of Marine Programs, 2004.

Chapter 6. Creepy Crawlies: Insects and Spiders

Pfeiffer, William J., and R. G. Wiegert. "Grazers on *Spartina* and Their Predators." In *The Ecology of a Salt Marsh*, edited by Lawrence R. Pomeroy and Richard G. Wiegert, 87–112. New York: Springer-Verlag, 1981.

Smalley, Alfred E. "Energy Flow of a Salt Marsh Grasshopper Population." *Ecology* 41 (1960): 672–77.

Chapter 7. Marsh Life: Scales

Crook, Ray. *A Place Known as Chocolate*. Carrollton: University of West Georgia, Antonio J. Waring Jr. Archaeological Laboratory, 2007. http://waring.westga.edu/Publication/Chocolate.pdf.

Dahlberg, Michael D., and Eugene P. Odum. "Annual Cycles of Species Occurrence, Abundance, and Diversity in Georgia Estuarine Fish Populations." *American Midland Naturalist* 83, no. 2 (1970): 382–92.

Dodd, C. Kenneth. *Synopsis of the Biological Data on the Loggerhead Sea Turtle "Caretta caretta" (Linnaeus 1758)*. Biological Report 88(14). Washington, D.C.: U.S. Fish and Wildlife Service, 1988. http://fl.biology.usgs.gov/Amphibians_and_Reptiles/LoggerheadSynopsis1988.pdf.

Ernst, Carl H., and Jeffrey E. Lovich. *Turtles of the United States and Canada*. 2nd ed. Baltimore: Johns Hopkins University Press, 2009.

Johnson, A. Sydney, Hilburn O. Hillestad, Sheryl F. Shanholtzer, and G. Frederick Shanholtzer. *An Ecological Survey of the Coastal Region of Georgia*. National Park Service Scientific Monograph 3, 1974.

Kneib, Ronald T. "Predation Risk and Use of Intertidal Habitats by Young Fishes and Shrimp." *Ecology* 68 (1987): 379–86.

———. "Seasonal Abundance, Distribution and Growth of Postlarval and Juvenile Grass Shrimp (*Palaemonetes pugio*) in a Georgia, USA, Salt Marsh." *Marine Biology* 96 (1987): 215–23.

Martof, Bernard S. "Some Observations on the Herpetofauna of Sapelo Island, Georgia." *Herpetologica* 19, no. 1 (1963): 70–72.

Teal, Mildred, and John Teal. *Portrait of an Island*. Athens: University of Georgia Press, 1967. Reprint, 1997.

Chapter 8. Marsh Life: Feathers and Fur

Beaton, Giff P., Paul W. Sykes Jr., and John W. Parrish Jr. *Annotated Checklist of Georgia Birds*. Occasional Publication 14. Athens: Georgia Ornithological Society, 2003.

Bond, B. T., M. I. Nelson, and R. J. Warren. "Home Range Dynamics and Den Use of Nine-Banded Armadillos on Cumberland Island, Georgia." *Proceedings of the Annual Conference of the Southeastern Association of Fish and Wildlife Agencies* 54 (2000): 415–23.

Johnson, A. Sydney, Hilburn O. Hillestad, Sheryl F. Shanholtzer, and G. Frederick Shanholtzer. *An Ecological Survey of the Coastal Region of Georgia.* National Park Service Scientific Monograph 3, 1974.

Kale, Herbert W. *Ecology and Bioenergetics of the Long-Billed Marsh Wren, "Telmatodytes palustris griseus," in a Salt Marsh Ecosystem.* Cambridge, Mass.: Publications of the Nuttall Ornithological Club, 1965.

Meanley, Brooke. *The Marsh Hen: A Natural History of the Clapper Rail of the Atlantic Coast Salt Marsh.* Centreville, Md.: Cornell Maritime Press / Tidewater Publishers, 1985.

Sharp, Homer F. "Food Ecology of the Rice Rat, *Oryzomys palustris* (Harlan), in a Georgia Salt Marsh." *Journal of Mammalogy* 48, no. 4 (1967): 557–63.

Teal, John M. "Birds of Sapelo Island and Vicinity." *Oriole* 24 (1959): 1–14, 17–20.

Chapter 9. What Lies Beneath: Zooplankton

Bentlage, Bastian, Paulyn Cartwright, Angel A. Yanagihara, Cheryl Lewis, Gemma S. Richards, and Allen G. Collins. "Evolution of Box Jellyfish (Cnidaria: Cubozoa), a Group of Highly Toxic Invertebrates." *Proceedings of the Royal Society B* 277 (2010): 493–501.

Brinkman, Diane L., and James N. Burnell. "Biochemical and Molecular Characterisation of Cubozoan Protein Toxins." *Toxicon* 54, no. 8 (2009): 1162–73.

Brotz, Lucas, William W. L. Cheung, Kristin Kleisner, Evgeny Pakhomov, and Daniel Pauly. "Increasing Jellyfish Populations: Trends in Large Marine Ecosystems." *Hydrobiologia* 690 (2012): 3–20.

Cartwright, Paulyn, Susan L. Halgedahl, Jonathan R. Hendricks, Richard D. Jarrard, Antonio C. Marques, Allen G. Collins, and Bruce S. Lieberman. "Exceptionally Preserved Jellyfishes from the Middle Cambrian." *PLoS ONE* 2, no. 10 (2007): e1121. doi:10.1371/journal.pone.0001121.

Condon, Robert H., Carlos M. Duarte, Kylie A. Pitt, Kelly L. Robinson, Cathy H. Lucas, Kelly R. Sutherland, Hermes W. Mianzan, Molly Bogeberg, Jennifer E. Purcell, Mary Beth Decker, Shin-ichi Uye, Laurence P. Madin, Richard D. Brodeur, Steven H. D. Haddock,

Alenka Malej, Gregory D. Parry, Elena Eriksen, Javier Quiñones, Marcelo Acha, Michel Harvey, James M. Arthur, and William M. Graham. "Recurrent Jellyfish Blooms Are a Consequence of Global Oscillations." *Proceedings of the National Academy of Sciences (USA)* 110, no. 3 (2013): 1000–1005.

Gershwin, Lisa-ann. *Stung! On Jellyfish Blooms and the Future of the Ocean.* Chicago: University of Chicago Press, 2013.

Purcell, Jennifer E. "Jellyfish and Ctenophore Blooms Coincide with Human Proliferations and Environmental Perturbations." *Annual Review of Marine Science* 4 (2012): 209–35.

Robertson, J. Roy. "Predation by Estuarine Zooplankton on Tintinnid Ciliates." *Estuarine, Coastal and Shelf Science* 16, no. 1 (1983): 27–36.

Smith, DeBoyd L. *A Guide to Marine Coastal Plankton and Marine Invertebrate Larvae.* Dubuque, Iowa: Kendall/Hunt, 1977.

Uye, Shin-Ichi. "Blooms of the Giant Jellyfish *Nemopilema nomurai*: A Threat to the Fisheries Sustainability of the East Asian Marginal Seas." *Plankton and Benthos Research* 3, suppl. (2008): 125–31.

Webb, Stacey, and Ronald T. Kneib. "Individual Growth Rates and Movement of Juvenile White Shrimp (*Litopenaeus setiferus*) in a Tidal Marsh Nursery." *Fishery Bulletin* 102, no. 2 (2004): 376–88.

Chapter 10. Attachment to Place: Settlers

"A Guide to Benthic Invertebrates and Cryptic Fishes of Gray's Reef." www.bio.georgiasouthern.edu/GR-inverts.

Johnson, A. Sydney, Hilburn O. Hillestad, Sheryl F. Shanholtzer, and G. Frederick Shanholtzer. *An Ecological Survey of the Coastal Region of Georgia.* National Park Service Scientific Monograph 3, 1974.

Chapter 11. Sound Swimmers: Nekton

Associated Press. News Archive, Sports Shorts, August 1, 1985. http://www.apnewsarchive.com/1985/Sports-Shorts/id-df711a7e1f92e4a24e7a2348ef05daf6.

Dahlberg, Michael D. *Guide to Coastal Fishes of Georgia and Nearby States.* Athens: University of Georgia Press, 1975.

Dahlberg, Michael D., and Eugene P. Odum. "Annual Cycles of Species Occurrence, Abundance, and Diversity in Georgia Estuarine Fish Populations." *American Midland Naturalist* 83, no. 2 (1970): 382–92.

Dresser, Brian K., and Ronald T. Kneib. "Site Fidelity and Movement Patterns of Wild Subadult Red Drum, *Sciaenops ocellatus* (Linnaeus), within a Salt Marsh–Dominated Estuarine Landscape." *Fisheries Management and Ecology* 14, no. 3 (2007): 183–90.

Franklin, H. Bruce. *The Most Important Fish in the Sea: Menhaden and America*. Washington, D.C.: Island Press, 2007.

Harvey, Chris J. "Use of Sandy Beach Habitat by *Fundulus majalis*, a Surf-Zone Fish." *Marine Ecology Progress Series* 164 (1998): 307–10.

Hoese, H. D. "Dolphin Feeding out of Water in a Salt Marsh." *Journal of Mammalogy* 52, no. 1 (1971): 222–23.

Johnson, A. Sydney, Hilburn O. Hillestad, Sheryl F. Shanholtzer, and G. Frederick Shanholtzer. *An Ecological Survey of the Coastal Region of Georgia*. National Park Service Scientific Monograph 3, 1974.

Kneib, Ronald T. "The Role of *Fundulus heteroclitus* in Salt Marsh Trophic Dynamics." *American Zoologist* 26 (1986): 259–69.

———. "Size-Specific Patterns in the Reproductive Cycle of the Killifish, *Fundulus heteroclitus* (Pisces: Fundulidae) from Sapelo Island, Georgia." *Copeia* 2 (1986): 342–51.

Odum, William E. "The Ecological Significance of Fine Particle Selection by the Striped Mullet *Mugil cephalus*." *Limnology and Oceanography* 13, no. 1 (1968): 92–98.

Rickards, William L. "Ecology and Growth of Juvenile Tarpon, *Megalops atlanticus*, in a Georgia Salt Marsh." *Bulletin of Marine Science* 18, no. 1 (1968): 220–39.

Chapter 12. On, and under, the Beach: Living in Sand

Dörjes, Jürgen, Robert W. Frey, and James D. Howard. "Origins of, and Mechanisms for, Mollusk Shell Accumulations on Georgia Beaches." *Senckenbergiana Maritima* 18, nos. 1–2 (1986): 1–43.

Frey, Robert W. "Distribution of Ark Shells (Bivalvia: Anadara), Cabretta Island Beach, Georgia." *Southeastern Geology* 27 (1987): 155–63.

———. "Hermit Crabs: Neglected Factors in Taphonomy and Palaeoecology." *Palaios* 2 (1987): 313–22.

Johnson, A. Sydney, Hilburn O. Hillestad, Sheryl F. Shanholtzer, and G. Frederick Shanholtzer. *An Ecological Survey of the Coastal Region of Georgia*. National Park Service Scientific Monograph 3, 1974.

South Carolina State Parks. "Myrtle Beach State Park Beachcombing Guide." http://www.southcarolinaparks.com/files/State%20Parks%20Files/Myrtle%20Beach/MBSP-Beachcombers-Guide.pdf.

Telford, Malcolm. "A Hydrodynamic Interpretation of Sand Dollar Morphology." *Bulletin of Marine Science* 31, no. 3 (1981): 605–22.

Chapter 13. Loggerheads

Conant, Therese A., Peter H. Dutton, Tomoharu Eguchi, Sheryan P. Epperly, Christina C. Fahy, Matthew H. Godfrey, Sandra L. MacPherson, Earl E. Possardt, Barbara A. Schroeder, Jeffrey A. Seminoff, Melissa L. Snover, Carrie M. Upite, and Blair E. Witherington. "Loggerhead Sea Turtle (*Caretta caretta*) 2009 Status Review under the U.S. Endangered Species Act." Report of the Loggerhead Biological Review Team to the National Marine Fisheries Service, August 2009. http://www.nmfs.noaa.gov/pr/pdfs/statusreviews/loggerheadturtle2009.pdf.

Plotkin, Pamela T., and James R. Spotila. "Post-nesting Migrations of Loggerhead Turtles *Caretta caretta* from Georgia, USA: Conservation Implications for a Genetically Distinct Subpopulation." *Oryx* 36, no. 4 (2002): 396–99.

Chapter 14. Shorebirds

American Birding Association. "Identification of North American Peeps: A Different Approach to an Old Problem." *Birding*, July–August 2008, 32–40. https://www.aba.org/birding/v40n4p32.pdf.

Teal, John M. "Birds of Sapelo Island and Vicinity." *Oriole* 24 (1959): 1–14, 17–20.

Wilson, Jim. *Common Birds of Coastal Georgia*. Athens: University of Georgia Press, 2011.

Chapter 15. Seasons in the Sun

Bailey, Cornelia W., and Christena W. Bledsoe. *God, Dr. Buzzard, and the Bolito Man: A Saltwater Geechee Talks about Life on Sapelo Island*. New York: Doubleday, 2000.

Fry, Brian, and Evelyn B. Sherr. "$\delta^{13}C$ Measurements as Indicators of Carbon Flow in Marine and Freshwater Ecosystems." *Contributions in Marine Science* 27 (1984): 13–47.

Giurgevich, John R., and E. Lloyd Dunn. "Seasonal Patterns of CO_2 and Water Vapor Exchange of the Tall and Short Height Forms of *Spartina alterniflora* Loisel in a Georgia Salt Marsh." *Oecologia* 43 (1979): 139–56.

Haines, Evelyn B. "Stable Carbon Isotope Ratios in the Biota, Soils, and Tidal Water of a Georgia Salt Marsh." *Estuarine and Coastal Marine Science* 4 (1976): 609–19.

Teal, Mildred, and John Teal. *Portrait of an Island*. Athens: University of Georgia Press, 1967. Reprint, 1997.

Chapter 16. The Once and Future Coast

Balmer, Brian C., Lori H. Schwacke, Randall S. Wells, R. Clay George, Jennifer Hoguet, John R. Kucklick, Suzanne M. Lane, Anthony Martinez, William A. McLellan, Patricia E. Rosel, Teri K. Rowles, Kate Sparks, Todd Speakman, Eric S. Zolman, and D. Ann Pabst. "Relationship between Persistent Organic Pollutants (POPs) and Ranging Patterns in Common Bottlenose Dolphins (*Tursiops truncatus*) from Coastal Georgia, USA." *Science of the Total Environment* 409 (2011): 2094–101.

Balmer, Brian C., Lori H. Schwacke, Randall S. Wells, Jeffrey D. Adams, R. Clay George, Suzanne M. Lane, William A. McLellan, Patricia E. Rosel, Kate Sparks, Todd Speakman, Eric S. Zolman, and D. Ann Pabst. "Comparison of Abundance and Habitat Usage for Common Bottlenose Dolphins between Sites Exposed to Differential Anthropogenic Stressors within the Estuaries of Southern Georgia, U.S.A." *Marine Mammal Science* 29, no. 2 (2013): E114–E135.

Barbier, Edward B., Sally D. Hacker, Chris Kennedy, Evamaria W. Koch, Adrian C. Stier, and Brian R. Silliman. "The Value of Estuarine and Coastal Ecosystem Services." *Ecological Monographs* 81 (2011): 169–93. http://dx.doi.org/10.1890/10-1510.1.

Bertness, Mark D., and Tyler C. Coverdale. "An Invasive Species Facilitates the Recovery of Salt Marsh Ecosystems on Cape Cod." *Ecology* 94 (2013): 1937–43. http://dx.doi.org/10.1890/12-2150.1.

Bertness, Mark, Brian Silliman, and Robert Jeffries. "Salt Marshes under Siege: Agricultural Practices, Land Development and Overharvesting of the Seas Explain Complex Ecological Cascades That Threaten Our Shorelines." *American Scientist* 92 (2004): 54–61.

Church, John A., Jonathan M. Gregory, Neil J. White, Skye M. Platten, and Jerry X. Mitrovica. "Understanding and Projecting Sea Level Change." *Oceanography* 24, no. 2 (2011): 130–43.

Craft, Christopher, Jonathan Clough, Jeff Ehman, Samantha Joye, Richard Park, Steve Pennings, Hongyu Guo, and Megan Machmuller. "Forecasting the Effects of Accelerated Sea-Level Rise on Tidal Marsh Ecosystem Services." *Frontiers in Ecology and the Environment* 7 (2009): 73–78.

Feagin, Rusty A., Douglas J. Sherman, and William E. Grant. "Coastal Erosion, Global Sea-Level Rise, and the Loss of Sand Dune Plant Habitats." *Frontiers in Ecology and the Environment* 3, no. 7 (2005): 359–64.

Feagin, Rusty A., William K. Smith, Norbert P. Psuty, Donald R. Young, M. Luisa Martínez, Gregory A. Carter, Kelly L. Lucas, James C. Gibeaut, Jane N. Gemma, and Richard E. Koske. "Barrier Islands: Coupling Anthropogenic Stability with Ecological Sustainability." *Journal of Coastal Research* 26, no. 6 (2010): 987–92.

Francis, Jennifer A., Weihan Chan, Daniel J. Leathers, James R. Miller, and Dana E. Veron. "Winter Northern Hemisphere Weather Patterns Remember Summer Arctic Sea-Ice Extent." *Geophysical Research Letters* 36 (2009): L07503. doi:10.1029/2009GL037274.

Francis, Jennifer A., and Stephen J. Vavrus. "Evidence Linking Arctic Amplification to Extreme Weather in Mid-Latitudes." *Geophysical Research Letters* 39, no. 6 (2012): L06801. doi:10.1029/2012GL051000.

Gedan, Keryn B., Matthew L. Kirwan, Eric Wolanski, Edward B. Barbier, and Brian R. Silliman. "The Present and Future Role of Coastal Wetland Vegetation in Protecting Shorelines: Answering Recent Challenges to the Paradigm." *Climatic Change* 106 (2011): 7–29. doi:10.1007/s10584-010-0003-7.

Intergovernmental Panel on Climate Change. "Climate Change 2013: The Physical Science Basis." Fifth Annual Assessment Report, November 2013. https://www.ipcc.ch/report/ar5/wg1.

Kintisch, Eli. "Can Coastal Marshes Rise above It All?" *Science* 341 (August 2, 2013): 480–81.

Kirwan, Matthew L., Glenn R. Guntenspergen, Andrea D'Alpaos, James T. Morris, Simon M. Mudd, and Stijn Temmerman. "Limits on the Adaptability of Coastal Marshes to Rising Sea Level." *Geophysical Research Letters* 37 (2010): L23401-5. doi:10.1029/2010GL045489.

Kirwan, Matthew L., and J. Patrick Megonigal. "Tidal Wetland Stability in the Face of Human Impacts and Sea-Level Rise." *Nature* 504 (December 2013): 53–60.

Kleppel, Gary S., M. Richard DeVoe, and Mac V. Rawson, eds. *Changing Land Use Patterns in the Coastal Zone*. New York: Springer, 2006.

Lee, Richard F., and Keith A. Maruya. "Chemical Contaminants Entering Estuaries in the South Atlantic Bight as a Result of Current and Past Land Use." In *Changing Land Use Patterns in the Coastal Zone*, edited by Gary S. Kleppel, M. Richard DeVoe, and Mac V. Rawson, 205–27. New York: Springer, 2006.

Nicholls, Robert J. "Planning for the Impacts of Sea Level Rise." *Oceanography* 24, no. 2 (2011): 144–57. http://dx.doi.org/10.5670/oceanog.2011.34.

Ogburn, Matthew B., and Merryl Alber. "An Investigation of Salt Marsh Dieback in Georgia Using Field Transplants." *Estuaries and Coasts* 29, no. 1 (2006): 54–62.

Osgood, David T., and Brian R. Silliman. "From Climate Change to Snails: Potential Causes of Salt Marsh Die-Back along the U.S. Eastern Seaboard and Gulf Coasts." In *Human Impacts on Salt Marshes: A Global Perspective*, edited by Brian R. Silliman, Mark D. Bertness, and Edwin D. Grosholz, 231–52. Berkeley and Los Angeles: University of California Press, 2009.

Petes, Laura E., Alicia J. Brown, and Carley R. Knight. "Impacts of Upstream Drought and Water Withdrawals on the Health and Survival of Downstream Estuarine Oyster Populations." *Ecology and Evolution* 2, no. 7 (2012): 1712–24.

Pilkey, Orrin, and Rob Young. *The Rising Sea*. Washington, D.C.: Island Press, 2009.

Schaeffer, Michiel, William Hare, Stefan Rahmstorf, and Martin Vermeer. "Long-Term Sea-Level Rise Implied by 1.5°C and 2°C Warming Levels." *Nature Climate Change* 2 (2012): 867–70.

Silliman, Brian R., Edwin D. Grosholz, and Mark D. Bertness. "Salt Marshes under Global Siege." In *Human Impacts on Salt Marshes: A Global Perspective*, edited by Brian R. Silliman, Mark D. Bertness, and Edwin D. Grosholz, 391–98. Berkeley and Los Angeles: University of California Press, 2009.

Silliman, Brian R., and Steven Y. Newell. "Fungal Farming in a Snail." *Proceedings of the National Academy of Sciences (USA)* 100, no. 26 (2003): 15643–48.

Silliman, Brian R., Johan van de Koppel, Mark D. Bertness, Lee E. Stanton, and Irving A. Mendelssohn. "Drought, Snails, and Large-Scale Die-Off of Southern U.S. Salt Marshes." *Science* 310, no. 5755 (2005): 1803–6.

Stanford, Jennifer D., Rebecca Hemingway, Eelco J. Rohling, Peter G. Challenor, Martin Medina-Elizalde, and Adrian J. Lester. "Sea-Level Probability for the Last Deglaciation: A Statistical Analysis of Far-Field Records." *Global and Planetary Change* 79, nos. 3–4 (2011): 193–203.

Stevenson, J. Court, and Michael S. Kearney. "Impacts of Global Climate Change and Sea-Level Rise on Tidal Wetlands." In *Human Impacts on Salt Marshes: A Global Perspective*, edited by Brian R. Silliman, Mark D. Bertness, and Edwin D. Grosholz, 171–206. Berkeley and Los Angeles: University of California Press, 2009.

Tang, Qiuhong, Xuejun Zhang, Xiaohua Yang, and Jennifer A. Francis. "Cold Winter Extremes in Northern Continents Linked to Arctic Sea Ice Loss." *Environmental Research Letters* 8, no. 1 (2013). doi:10.1088/1748-9326/8/1/014036.

Teal, John, and Mildred Teal. *Life and Death of the Salt Marsh*. New York: Ballantine Books, 1969.

Thoresen, Merrilee, Merryl Alber, and Randal L. Walker. "Trends in Recruitment and *Perkinsus marinus* Parasitism in the Eastern Oyster, *Crassostrea virginica*, within the Sapelo Island National Estuarine Research Reserve (SINERR)." Marine Technical Report Number 05-1. Athens: University of Georgia School of Marine Programs, 2004. https://georgiaseagrant.uga.edu/images/uploads/media/TR05-1.pdf.

Verity, Peter G. "A Decade of Change in the Skidaway River Estuary. I. Hydrography and Nutrients." *Estuaries* 25, no. 5 (2002): 944–60.

———. "A Decade of Change in the Skidaway River Estuary. II. Particulate Organic Carbon, Nitrogen, and Chlorophyll a." *Estuaries* 25, no. 5 (2002): 961–75.

Verity, Peter G., Merryl Alber, and Suzanne B. Bricker. "Development of Hypoxia in Well-Mixed Subtropical Estuaries in the Southeastern USA." *Estuaries and Coasts* 29, no. 4 (2006): 665–73.

Verity, Peter G., and David G. Borkman. "A Decade of Change in the Skidaway River Estuary. III. Plankton." *Estuaries and Coasts* 33 (2010): 513–40.

Zhang, Keqi, Bruce C. Douglas, and Stephen P. Leatherman. "Global Warming and Coastal Erosion." *Climatic Change* 64 (2004): 41–58.

Appendix 1. Where to Go to Enjoy Georgia Coastal Wildlife

Georgia Department of Natural Resources. "Georgia's Colonial Coast Birding Trail." http://georgiawildlife.com/ColonialCoastBirdingTrail.

Appendix 3. Methods for Collecting and Inspecting Coastal Biota

Williams, Richard B. "Use of Netting to Collect Motile Benthic Algae." *Limnology and Oceanography* 8 (1963): 360–61.

INDEX

Abra aequalis (small estuarine clam), 136
Acanthohaustorius (intertidal beach amphipod), 138
Acartia tonsa (coastal copepod), 107
acorn worm (*Saccoglossus kowalevskii*), 49, 51, 55
Actitis macularius (spotted sandpiper), 153
Aedes sollicitans (salt marsh mosquito), 77
Aedes taeniorhynchus (salt marsh mosquito), 77
Agelaius phoeniceus (red-winged blackbird), 88–89
Agkistrodon piscivorus (eastern cottonmouth snake), 82, 84
Agraulis vanillae (Gulf fritillary butterfly), 78
Alcyonidium hauffi (dead man's fingers bryozoan), 111, 112–13
algae: crabs feeding on, 68–69, 70, 71; diatoms, 26–27, 28–29, 56, 59, 107, 115; filamentous, 29, 31; in fouling communities, 65; insects feeding on, 76–77; large (macrophytic), 6, 10, 25, 29–30; in lichens, 32–33; meiofauna feeding on, 51, 52; microscopic, 6, 10, 19, 20, 22, 24–29; mollusks feeding on, 59, 60; phytoflagellates, 27–28, 29, 30; polychaetes feeding on, 54; production of, 166, 169; role in estuarine food web, 166, 169–70; role in marine ecosystems, 26–27; zooplankton feeding on, 13, 101. *See also* benthic algae; diatoms; dinoflagellates; euglenae
algal blooms, 26–27, 29, 30, 31, 178
Alitta succinea (common clam worm), 54
Alligator mississippiensis (American alligator), 2, 6, 82, 163, 178, 195
Alosa sapidissima (American shad), 121, 123–24
Alpheus heterochaelis (snapping shrimp), 67, 125
Altamaha River, xiv, 38, 93, 105, 130, 151, 195

Americorchestia longicornis (beach flea amphipod), 65, 138, 163
Ammodramus maritimus (seaside sparrow), 88
Ammospiza caudacuta (sharp-tail sparrow), 88
amoebae, 20, 23, 24
amphibians, 49, 84, 90
amphipods, 7, 65, 77, 109, 138, 153
anchovies, 121, 123
anemones, 56–57, 102, 110, 111–12, 146
Anguilla rostrata (American eel), 129
Anguinella palmata (ambiguous bryozoan), 112
anhinga (*Anhingha anhinga*), 154
annelid worms, 53. *See also* bristle worms
Anolis carolinensis (green anole lizard), 83–84
anomuran crabs, 68, 140–41. *See also* crabs; hermit crabs
ant lions (*Glenurus*), 77
ants, 74
Aplidium stellatum (sea pork tunicate), 114
Arabella iricolor (opal polychaete worm), 54
archaea, 15–16, 17
Archosargus probatocephalus (sheepshead fish), 121, 126
Ardea alba (great egret), 91
Ardea herodias (great blue heron), 6, 89–90
Arenaeus cribrarius (speckled crab), 120
Arenaria interpres (ruddy turnstone), 152
Ariopsis felis (sea catfish), 121, 128
armadillo (*Dasypus novemcinctus*), 99–100
Armases cinereum (wharf crab), 71, 162
Arthropoda, 52, 58, 118
Asterias forbesi (common sea star), 143–44
Asterionellopsis glacialis (planktonic chain-forming diatom), 27

Astropecten articulatus (margined sea star), 144
Atrina serrata (saw-toothed pen shell), 137
Atriplex cristata (saltbush or orach), 46
Austinixa cristata (commensal pea crab), 139
autotrophs, 19, 24–25
Aythya affinis (lesser scaup duck), 155

Baccharis halimifolia (groundsel bush), 6, 41, 89
bacteria, 15–19; cordgrass decomposition by, 11; cyanobacteria, 18–19, 24, 25, 32, 199; fermenters, 16; filamentous, 17–19; marine viruses and, 15; in marsh mud, 51, 52; meiofauna feeding on, 51, 52; respiration rates, 12; role in marine ecosystems, 10, 13–14, 23–24; sulfate-reducing, 16–18, 36; sulfide-oxidizing, 17–18, 24
Bagre marinus (gafftopsail catfish), 128
Bailey, Cornelia, 157–58, 194
Bairdiella chrysoura (silver perch), 122, 124
Balanus (acorn barnacle), 113
bald eagle (*Haliaeetus leucocephalus*), 95, 193, 194
barnacles, 25, 60, 109, 113
basket making, 44–45
Batis maritima (saltwort), 4, 40, 159, 161
Bdelloura candida (horseshoe crab leech), 142
beach and dune habitat, 1–2, 7–9, 44–48, 77–78, 188. *See also* sand-dwelling creatures
beach flea amphipods, 65, 138, 163
belted kingfisher (*Megaceryle alcyon*), 93–94
benthic algae: abundance of, 7, 28, 29; blooms of, 29, 199; carbon isotope analysis, 168; collection methods, 199; as food source, 25, 29, 53, 56, 59, 170; pennate diatoms, 27, 28–29; production, 166
benthic animals: anemones, 102; bryozoans, 112–13; clams, 133–34; collection methods, 202; filter-feeders, 26; harpacticoid copepods, 53; haustoriid amphipods, 138; meiofauna, 51–52, 59; number of, 6–7; oxygen scavenging, 54; as prey, 88, 148; sponges, 109–10; tanaids, 55–56
Biffarius biformis (ghost shrimp), 139–40
biogeochemical cycling, 18
bioluminescence, 21, 107, 111
birds, 86–96; diversity of species on Georgia coast, 87; evolution of, 86–87; poisoned by pesticides, 178; as predators, 24, 50, 71, 73, 75, 79, 94–95, 123, 148, 162, 166; as prey, 84, 94, 98; songbirds, 89, 194; wading, 79, 194; in winter, 164. *See also* shorebirds; *and specific bird species*
bivalve mollusks, 58, 60–65, 135–37; as prey, 110, 134, 144, 176. *See also* mollusks
Blackbeard Island, xiv, 157–58
blackbirds, 88–89
black needlerush (*Juncus roemerianus*), 4–6, 38
blue crab (*Callinectes sapidus*): crab barnacle and, 113; at high tide, 64, 162; life history, 119–20; as nekton, 119–20; parasitic infection of, 182–84; periwinkle snail populations and, 184–85; as predator, 29, 64, 72, 79, 120, 162; as prey, 118; *Spartina* carbon in, 170
boat-tailed grackle (*Quiscalus major*), 88–89
boring sponges, 63, 110
Borrichia frutescens (sea oxeye bush), 40–41, 161
Bostrychia radicans (macrophyte alga), 25, 31
bottlenose dolphins (*Tursiops truncatus*), 6, 115, 123, 130–31, 179–80, 193
Bougainvillia carolinensis (fouling community hydroid), 110
brachyuran crabs, 68. *See also* crabs; decapods
Brevoortia tyrannus (menhaden or pogies), 6, 123
bristle worms (polychaetes), 7, 53–55; commensal crabs in burrows of, 142; larvae of, 108; macrofaunal, 53–55; meiofaunal, 52; as prey, 79
brittle stars, 63, 144
bryozoans (moss animals), 6, 112–13; amphipods among, 65; on dead sea whips, 111; on floating docks, 109; in oyster reefs, 63; spider crabs growing with, 142
Bubulcus ibis (cattle egret), 74, 92–93
Bugula (tufted bryozoan), 112
bumper, Atlantic (*Chloroscombrus chrysurus*), 129
bushes, 6; marsh edge, 32, 40–41; dune, 43, 46–47
Busycon carica (knobbed whelk), 132, 133, 140
Busycon contrarium (lightning whelk), 132
Busycotypus canaliculatus (channeled whelk), 133
Butorides virescens (green heron), 90–91
butterflies, 78

c-3 pathway of photosynthesis, 34, 160, 168
c-4 pathway of photosynthesis, 34, 35, 39, 159–60, 168
cactus, 43, 45–46
Cakile edentula (sea rocket herb), 46
calcium carbonate, 23, 58, 109–10, 117, 143, 167
Calidris alba (sanderling), 152
Calidris alpina pacifica (dunlin), 152
Calidris canutus rufa (red knot), 152
Calidris maritima (purple sandpiper), 152
Calidris mauri (western sandpiper), 152
Calidris minutilla (least sandpiper), 152
Calidris pusilla (semipalmated sandpiper), 152
Calliactis tricolor (tricolor anemone), 111
Callichirus major (ghost shrimp), 7, 67, 138–39
Callinectes sapidus (blue crab). *See* blue crab
Caloglossa leprieurii (filamentous red alga), 31
Cambrian period, 34, 52, 103
camphor weed (*Heterotheca subaxillaris*), 46
Caprella (skeleton shrimp), 65
carbon fixation, 15–16, 17–18, 19, 24, 34, 159–60
carbon isotopes and geochemistry, xiii, 166–69
Carcharhinus isodon (finetooth shark), 130
Carcharhinus plumbeus (sandbar shark), 130
Carcharias taurus (sand tiger shark), 130
Carcinus maenas (European green crab), 176
Caretta caretta (loggerhead turtle), 8, 81–82, 146–49, 195
catfish, 121, 127–28
cattle egret (*Bubulcus ibis*), 74, 92–93
Cemophora coccinea (scarlet snake), 83
Cenchrus tribuloides (dune sandspur), 48
centric diatoms, 27, 28
Centropristis striata (black sea bass), 121, 126
cephalopods, 117–18, 167
Ceratium (armored dinoflagellate with horns), 21
cetaceans (whale family), 115, 130–31
chachalaca pheasant (*Ortalis vetula*), 95–96
Chaetoceros (planktonic chain-forming diatom with spines), 27
Chaetodipterus faber (Atlantic spadefish), 129
Chaetognatha (arrow worms), 108
Charadrius semipalmatus (semipalmated plover), 152

Charadrius vociferus (killdeer), 152
Charadrius wilsonia (Wilson's plover), 152
Chelonibia patula (crab barnacle), 113
chemosynthesis, 17–18
Chesapeake Bay, 62, 104, 126, 185
chloroplasts, 19, 20, 21, 24–25
Chloroscombrus chrysurus (Atlantic bumper), 129
Chroicocephalus philadelphia (Bonaparte's gull), 150
chromatophores, 117–18
Chrysaora quinquecirrha (sea nettle jellyfish), 104–5
Chrysops (deer fly), 76
Chthamalus fragilis (little gray barnacle), 113
Cicindela (tiger beetle), 77
ciliates, 20, 22–23, 24, 51
Circus cyaneus (harrier hawk), 94
Cistothorus palustris (long-billed marsh wren), 87–88
clams, 6, 7; as filter feeders, 6, 26, 136; mud-dwelling, 64–65; as prey, 7, 29, 83, 98, 133–34, 135, 146, 152; sand- or surf-dwelling, 135–37; as shellfish, 58. *See also* bivalve mollusks; *and specific clam species*
clapper rail (*Rallus longirostris*), 6, 64, 86, 87, 156
Clathria prolifera (red beard sponge), 109–10
Clibanarius vittatus (thinstripe hermit crab), 140
climate change, 174, 180–86
Cliona (shell-burrowing sponge that degrades oyster shells), 110
Clubiona littoralis (leafcurling sac spider), 74
Clymenella torquata (bamboo worm, polychaete), 55
Cnidaria, 101–3
Coffin, Howard, ix–x, 62, 96
Colonial Coast Birding Trail, 87, 193
Coluber constrictor (black racer snake), 83
comb jellies (Ctenophora), 106–7, 201
composite plants, 40–41, 46
conservation organizations, 178–89, 197–98
copepods, 107–8; calanoid, 53, 107, 118; food sources for, 26; harpacticoid, 53, 80, 138; observing collected, 201; as prey, 56, 106, 110, 115
coquina surf clam (*Donax variabilis*), 135–36, 152–53

Coragyps atratus (black vulture), 95
corals, 6, 102, 110–11
cordgrass *(Spartina alterniflora)*, 3–4, 5;
    adaptations to environment, 159–61;
    algae colonization of, 10, 29, 31; bacterial
    decomposition of, 11, 23–24; creatures on
    leaves of, 53, 56, 162; creekside, 4, 5, 37, 38, 161,
    175–76; effect of sea-level rise on, 187–89; fungi
    growth on, 10, 20, 23–24; giant cordgrass (*S.
    cynosuroides*), 38; insects feeding on, 73–74;
    oxygen acquisition, 36; photosynthesis, 34,
    159–60, 161; production of, xii, xiii, 11, 165,
    169; rafts of stems, 37–38, 169; respiration, 161;
    salt excretion, 4, 35, 36, 161; salt marsh hay
    cordgrass (*S. patens*), 44, 97; salt tolerance,
    159, 161; seasonal colors, 32; snails on stalks of,
    58, 59, 181–82; sulfide effects on, 36–37; sulfur
    acquisition, 160
cordgrass-detritus food web, xii–xiii, 11, 13, 165–70.
    See also food webs
cordgrass marshes, 3–4; animal density, 164; birds
    and mammals in, 86, 87–88, 94, 96–97, 98, 155;
    blue crabs in, 64, 120, 162, 170, 184–85; browning
    or dieback of, 181–82; drainage ditches in, 76,
    175; fiddler crabs in, 68–72; insects and spiders
    in, 73–76; killifish in, 73, 79–80; mussels in, 63–
    64; shorebird egg-laying in, 153, 156; squareback
    crab destruction of, 175; tidal rhythms in,
    162–63. *See also* cordgrass
Corvus ossifragus (fish crow), 95
Coscinodiscus (centric diatom), 27, 28
cottonmouth snake, eastern (*Agkistrodon
    piscivorus*), 82, 84
crabs: anomuran and brachyuran crabs, 68, 140–
    41; in fouling communities, 25, 109; green crabs,
    176; horseshoe crabs, 6, 52, 113, 142–43; king
    crabs, 118; lady crabs, 120; larvae, 108; marsh
    crabs, 71, 175; mating, 70–71; mud-dwelling, 58,
    59, 63, 64, 68–72; near-shore, 7, 137; as nekton,
    115, 118; as predators, 24, 166; as prey, 6, 90,
    98, 146, 148; sand-dwelling, 140–43; *Spartina*
    carbon in, 169, 170; speckled crabs, 120; spider
    crabs, 105, 141–42; true crabs, 68, 140. *See also*
    blue crab; fiddler crabs; hermit crabs

Crassostrea virginica (American oyster), 60
creekside marsh, 4, 5, 36, 37
Crematogaster laeviuscula (acrobatic ant), 74
Crepidula fornicata (common slipper shell), 135
Crepidula plana (eastern white slipper shell), 135
Cretaceous period, 33, 82, 86–87, 167
croaker (*Micropogonias undulatus*), 121, 124,
    125, 126
Crotalus adamanteus (eastern diamondback
    rattlesnake), 8, 44, 82, 84–85
Croton punctatus (beach croton), 46
crows, 95
Cruciferae (mustard plant family), 46
crustaceans: amphipods, 7, 65, 77, 109, 138, 153;
    barnacles, 25, 60, 109, 113; decapods, 66–72,
    118–20; food sources for, 29; isopods, 55, 65–66;
    meiofaunal, 51, 52–53; mud-dwelling, 51, 52–53,
    55–56, 58, 65–72; as prey, 7, 79, 152, 176; sand-
    dwelling, 132, 137–43; in zooplankton, 13. *See
    also* copepods; crabs; shrimp
Ctenophora (comb jellies), 106–7, 201
Culicoides (sand gnat or no-see-um), 76
Cumberland Island (Cumberland Island National
    Seashore), xiv, 87, 157, 196
curlews, 153, 154
Cyanea capillata (lion's mane jellyfish), 104
cyanobacteria, 18–19, 24, 25, 32, 199
Cyathura burbancki (estuarine isopod), 66
Cyathura polita (estuarine isopod), 65–66
Cynoscion nebulosus (spotted sea trout), 124–25
Cynoscion regalis (weakfish or summer trout),
    125, 126
Cyprideis floridana (marsh mud ostracod), 53
Cyprinodon variegatus (sheepshead minnow), 81
Cyrtopleura costata (angel wing clam), 136

Dasyatidae (stingray), 129–30
Dasypus novemcinctus (nine-banded armadillo),
    99–100
dead man's fingers bryozoan (*Alcyonidium
    hauffi*), 111, 112–13
decapods, 66–72, 118–20
deer, white-tailed (*Odocoileus virginianus*), 2, 40,
    43, 86, 99

deer fly (*Chrysops*), 76
dermo oyster disease, 62, 185–86
development, coastal, 177–78, 189
Devonian period, 33–34
Diadumene leucolena (ghost anemone), 111
diatoms, 26–27, 28–29, 56, 59, 107, 115, 199
Dinocardium robustum (robust Atlantic cockle), 136
dinoflagellates, 20–22, 107, 164, 183–85
Diopatra cuprea (plumed polychaete worm), 55
Dissodactylus mellitae (sand-dollar crab), 142
Distichlis spicata (salt grass), 4, 39
Doboy Sound, 116, 120–21, 155, 158
dolphins, 6, 115, 123, 130–31, 179–80, 193
Donax variabilis (coquina surf clam), 135–36, 152–53
Doryteuthis pealeii (long-finned squid), 117
Dosinia discus (disk clam), 136
double-crested cormorant (*Phalacrocorax auritus*), 154
Drilonereis longa (threadworm polychaete), 54
droughts, 82, 130, 159, 174, 181–82, 184, 185–86
drum family of fishes, 124–26
ducks, 2, 150, 155–56, 164, 193, 195
dune plants, 44–48
Duplin River, viii, ix, 62, 94–95, 185

echinoderms, 108, 132, 143–45
ecosystem theory, 10–11, 13
Ectocarpus siliculosus (filamentous brown alga), 29
Ectopleura crocea (pink-hearted hydroid), 110
eel, American (*Anguilla rostrata*), 129
egrets, 74, 86, 90, 91–93, 163, 194. *See also* herons
Egretta caerulea (little blue heron), 90
Egretta rufescens (reddish egret), 90
Egretta thula (snowy egret), 91–92
Egretta tricolor (tricolored heron), 90
Elops saurus (ladyfish or ten-pounder), 124
Emerita talpoida (surf mole crab), 140–41, 152–53
Ensis directus (jackknife clam or razor clam), 137
Eontia ponderosa (ponderous ark), 136
estuaries, 1–7; animal abundance in, 49; climate change effects, 180–86; dead zones, 178; environmental stresses in, 158–59; pollution of, 175, 176–80; rafts of cordgrass stems in, 38; role of microbes in, 11, 16–19, 23–24, 25–26, 28; seasonal cycle, 163–64; subtidal estuary of Georgia coast, 2, 6–7, 101; tidal flooding of, 2, 11. *See also* food webs; salt marsh
Eubalaena glacialis (right whale), 131
Eudocimus albus (white ibis), 93, 193
euglenae, 28, 29, 30, 199
eukaryotic microbes, 19–31; algae, 24–31; defined, 19; fungi, 19–20; predatory protists, 20–23
Euphorbia polygonifolia (beach spurge herb), 46
Eurypanopeus depressus (black-fingered mud crab), 71
Eurytium limosum (broadback mud crab), 71
exoskeletons, 52, 58, 66, 118, 138

Farfantepenaeus aztecus (brown shrimp), 118–19
fertilizer run-off, 176, 177–78
fiddler crabs, 6, 7, 58, 68–71; active period, 163; burrows, 80; food sources for, 29, 50; marsh fiddlers, 29, 68–69, 164; as prey, 83, 87, 96, 98, 118, 125, 126, 151, 162; red-jointed fiddlers, 69; sand fiddlers, 39, 69, 126, 163; seasonal cycle, 163, 164; tidal rhythms and, 162; as true crabs, 140
fish, 7, 79–81; algae required for, 26–27, 29; bottom-feeding, 29; as chordates, 49; collection methods, 201–2; at high tide, 162; larval, 80, 104, 108; as nekton, 115, 120–30; overharvesting of, 103–4; parasitic infection of, 183; as predators, 24, 29, 56, 59, 63, 71–72, 79, 101, 107–8, 148, 166; as prey, 6, 80, 84, 90, 98, 104, 108, 146, 148, 150, 151, 154; seasonal cycle, 163; *Spartina* carbon in, 170. *See also specific fish species*
flagellates: choanoflagellates, 20; dinoflagellates, 20–22; disease-causing, 62; in marsh mud, 51; phosphate excretion, 11; phytoflagellates, 27–28, 29, 30; respiration rates, 11; role in marine ecosystems, 13–14, 22–23, 24
flatfish, 127
flies, 56, 73, 75, 76, 77–78
Florida pompano (*Trachinotus carolinus*), 126
flounder, 121, 127

food webs, 164–65; cordgrass-detritus food web, xii–xiii, 11, 13, 165–70; microbially enriched plant detritus and, 11, 13, 166; revised view of, 11–13, 169–170; role of algae and phytoplankton, 6, 26, 166, 168–69, 170; role of zooplankton, 6; Steele's model of, 12–13

foraminiferans, 23

fouling communities, 26, 63, 65, 66, 109–14, 142

freshwater marsh rabbit (*Sylvilagus aquaticus*), 97

frogs, 84, 90

Fundulus confluentus (marsh killifish), 80–81

Fundulus heteroclitus (mummichog killifish), 79–80, 94

Fundulus luciae (spotfin killifish), 80

Fundulus majalis (striped killifish), 81, 122

fungi, 10, 16, 19–20, 23–24, 32–33, 181–82

Gammarus mucronatus (marsh scud amphipod), 65

Gammarus palustris (marsh scud amphipod), 65

gastropods, 58–60, 111, 133–35, 144. *See also* mollusks

gelatinous zooplankton, 101–7

Gelochelidon nilotica (gull-billed tern), 151

geological time scale of evolution, 34

Georgia coast: animal density on, 164–65; climate and seasons, 2, 157–64; conservation organizations protecting, 197–98; geology of, 1–2; higher plants, 32–48; impacts on, 175–76; insects and spiders, 73–78; loggerhead turtles, 146–49; map of sea islands, xiv; marine habitats of, 1–9; marsh food webs, 164–70; microscopic life, 10–31; mud-dwelling invertebrates, 49–60, 63–65, 68–72; nekton, 115–31; oyster die-offs along, 185–86; pollution of, 176–80; sand-dwelling creatures, 132–45; sea-level rise and, 187–89; sessile organisms, 109–14; shorebirds, 150–56; vertebrates, 79–100; where to see wildlife, 193–96; zooplankton, 101–8

Geukensia demissa (ribbed mussel), 8–9, 63–64, 87, 162, 164

ghost crab (*Ocypode quadrata*), 7–8, 137, 141, 143, 149, 163

ghost shrimp, 7, 67, 138–40

giant cordgrass (*Spartina cynosuroides*), 38

Gibbesia neglecta (mantis shrimp), 68

gills: on burrowing polychaetes, 54, 55; on crabs, 68, 69; on horseshoe crabs, 142; on mollusks, 56, 59–60, 142

glassworts, 4, 40, 161

Glenurus (ant lion), 77

global warming, 174, 187, 188–89

Globicephala macrorhynchus (short-finned pilot whale), 131

Glycera (bloodworms), 54

gnats. *See* midges

gobies, 63, 81

Gobiosoma bosc (naked goby), 81

goldenrod, seashore (*Solidago sempervirens*), 46

Grammonota trivittata (sheet-web spider), 74

grasses, 16, 33–35. *See also* marsh grasses

grasshoppers, 74, 87, 93, 160

grass shrimp (*Paleomonetes pugio*), 56, 66–67, 80, 118, 124, 162

Gray's Reef (Gray's Reef National Marine Sanctuary), 7, 104, 109, 112

great blue heron (*Ardea herodias*), 6, 89–90

grebe, horned (*Podiceps auritus*), 155

Green, Allen, 44, 45

green beach algal blooms, 29, 30, 31

green crabs, European (*Carcinus maenas*), 176

greenhead fly (*Tabanus nigrovittatus*), 56, 76

gribbles (*Limnoria*), 66

groundsel bush (*Baccharis halimifolia*), 6, 41, 89

gulls (seagulls), 6, 8, 74, 150, 156

gymnosperms, 33, 41–42

Haematopus palliatus (American oystercatcher), 152

Haliaeetus leucocephalus (American bald eagle), 95

Halichondria bowerbanki (surface-attaching branched sponge), 110

hard-shell clam (*Mercenaria mercenaria*), 64–65

Hargeria rapax (marsh tanaid), 55–56

harpacticoid copepods, 53, 80, 138

Haustorius (beach sand amphipod), 138

hawks, 8, 94, 97

Hematodinium (crab parasite dinoflagellate), 183–84, 185
hemichordates, 49, 51, 55
Hemipholis elongata (banded brittle star), 144
Hepatus epheliticus (Dolly Varden, or calico, crab), 141
Hercules-club (*Zanthoxylum clava-herculis*), 42
hermit crabs: anemones on, 111; in beach sloughs, 122; as prey, 141; shells used by, 133, 140; slipper shells on, 135; snail fur on, 110
herons (including egrets), 2, 89–93; active periods, 6, 163; at Harris Neck National Wildlife Refuge, 194; as predators, 6, 74, 86; on Tolomato Island, 194–95
Heterotheca subaxillaris (camphor weed), 46
heterotrophs, 19, 24, 25
high tide, 3–4, 6, 7; animal activity at, 162, 163, 181, 184–85; cordgrass rafts carried in, 37, 38; diatoms at, 28; increasing height of, 186, 188
Hippocampus erectus (seahorse), 129
hogchoker flatfish (*Trinectes maculatus*), 127
Holocene epoch, 2
hooded merganser (*Lophodytes cucullatus*), 156
horseshoe crab, Atlantic (*Limulus polyphemus*), 6, 52, 113, 142–43
hurricanes, 157–58, 187
Hydractinia echinata (hydroid on hermit crab shells), 110
hydrobiid snails, 56, 59
Hydrocotyle bonariensis (pennywort), 48
hydroids, 6, 65, 106, 109, 110, 142
hydromedusae jellyfish, 106, 110, 201
hydrothermal vents, 18

ibis, 93, 193, 194–95
ice pack and sheet melting, 172, 173–74, 186, 187
Ilex vomitoria (yaupon bush), 42
Ilyanassa obsoleta (estuarine mud snail), 29, 59–60, 163, 164–65
indigo bunting (*Passerina cyanea*), 89
insects, 73–78; cordgrass as food for, 7, 162, 169; evolution of, 52; exoskeletons, 58; life cycle in salt marsh, 56; as prey, 79, 83, 87–88, 97, 151, 162. *See also specific insect species*

interdune meadows, 8, 81, 193
intertidal salt marsh, 2, 3–6
invertebrates, 49, 51–78. *See also* crustaceans; echinoderms; insects; mollusks; spiders; worms; zooplankton
iodine or acorn worm (*Saccoglossus kowalevskii*), 49, 51, 55
Ipomoea imperati (beach morning glory), 48
isopods, 55, 65–66
Iva frutescens (marsh elder bush), 6, 41
Iva imbricata (seashore elder bush), 46

Jekyll Island, xiv, 66, 149, 195–96
jellyfish, 101–6, 110, 117, 148, 201
Johannes, Robert, 11–12, 13
Juncus roemerianus (black needlerush), 4–6, 38
juniper (*Juniperus virginiana*), 6, 41–42
Jurassic period, 33–34, 87

keyhole urchin (*Mellita quinquiesperforata*), 142, 144
killifish, 79–81; at high tide, 71, 73, 162; at low tide, 67; in oyster reefs, 63, 67; as predators, 73; as prey, 67, 71, 92, 94, 120, 124; striped, 81, 122
Kinbergonuphis microcephala (burrowing polychaete worm), 54–55
kingfish (whiting, *Menticirrhus*), 121, 124
knobbed whelk (*Busycon carica*), 132, 133, 140
Kryptoperidinium (nontoxic red tide dinoflagellate), 21–22, 164

ladyfish (*Elops saurus*), 124
Laeonereis culveri (mud-burrowing marsh polychaete worm), 54
Lampropeltis getula (eastern king snake), 83
Lanier, Sidney, 35, 101
Larimus fasciatus (banded drum), 125–26
Larus sp. (gull or seagull), 150, 156
Leiostomus xanthurus (spot, small perch-like fish), 124, 126
Leptogorgia virgulata (sea whip), 111, 112–13
Leptosynapta inhaerens (worm sea cucumber), 145
Leucophaeus atricilla (laughing gull), 150, 156
Libinia dubia (spider crab), 141–42

Libinia emarginata (spider crab), 141–42
lichens, 32–33, 43
Life and Death of the Salt Marsh (J. Teal), xii, 175
Ligia exotica (exotic sea roach isopod), 66
Ligia oceanica (temperate sea roach isopod), 66
lilies, 45
Limnodromus griseus (short-billed dowitcher), 153
Limnoria (gribble, wood-boring isopod), 66
Limonium carolinianum (sea lavender), 41
limulus leech (*Bdelloura candida*), 142
Limulus polyphemus (Atlantic horseshoe crab), 6, 52, 113, 142–43
Lissodendoryx isodictyalis (garlic sponge), 110
Litopenaeus setiferus (white shrimp), 118–19
Littoraria irrorata. *See* marsh periwinkle snail
live oak (*Quercus virginiana*), ix, 6, 32, 42–44, 93, 96, 99
lizards, 82–84
loggerhead turtle (*Caretta caretta*), 8, 81–82, 146–49, 195
Lolliguncula brevis (brief thumbstall squid), 117
Lontra canadensis (river otter), 98–99
lookdown fish (*Selene vomer*), 129
Lophodytes cucullatus (hooded merganser), 156
lovebug fly (*Plecia nearctica*), 77–78
low marsh. *See* marsh plain
low tide, 3–4, 6, 7; animal activity at, 39, 162, 163; diatoms at, 28, 29; drought and evaporation at, 181; mudflat at, 50; plant desiccation at, 159; plants sheltering animals at, 41; rainstorms at, 166; water temperatures, 159
Lunarca ovalis (blood ark clam), 136

Macoma (tellin, thin-shelled burrowing clam), 136–37
macrofauna, 53–57
macrophyte algae, 25, 29–30, 31
Malaclemys terrapin (terrapin, marsh turtle), 83
mammals, 96–100; evolution of, 33–34, 87; marine, 115, 130–31; as predators, 71; as prey, 84. *See also* specific mammal species
Manayunkia aestuarina (salt marsh polychaete worm), 52
marsh browning or die-back, 181–82

"Marshes of Glynn, The" (Lanier), 35, 101
marsh grasses, 16, 27, 29, 34–40. *See also* cordgrass
marsh hen, 6. *See also* clapper rail (*Rallus longirostris*)
marsh mud, creatures of. *See* mud-dwelling creatures
marsh mud snails (hydrobiid), 56, 59
marsh periwinkle snail (*Littoraria irrorata*), 58, 59–60; blue crab populations and, 184–85; density in marsh, 164; food sources for, 29, 31; fungus farming, 181–82, 183; population outbreaks, 160; as prey, 83, 87, 120; during tidal cycle, 163; in winter, 163
marsh plain, 3, 4, 36, 38, 63, 68, 161, 165–66, 181, 185
marsh rabbit (*Sylvilagus*), 8, 97, 163
marsh rice rat (*Oryzomys palustris*), 86, 96–97, 163
marsh scud amphipods (*Gammarus*), 65
marsh wrens, 76, 87–88, 96–97
meadow marsh. *See* marsh plain
Megaceryle alcyon (belted kingfisher), 93–94
Megalops atlanticus (Atlantic tarpon), 115, 121, 124
meiofauna, 51–53, 56, 59, 69–70, 152, 202
Melampus bidentatus (marsh air-breathing snail), 60, 162
Melampus coffea (coffee bean snail), 60
Melanoleuca (mushroom-producing beach fungus), 20
Mellita quinquiesperforata (keyhole urchin), 142, 144
Membranipora (lacy crust bryozoan), 111, 112
Membras martinica (rough silverside), 122–23
menhaden (pogies, *Brevoortia tyrannus*), 6, 123
Menidia (Atlantic silverside), 122–23
Menticirrhus (kingfish or whiting), 121, 124
Mercenaria mercenaria (hard-shell or northern quahog clam), 64–65
methane, 17
mice, 8, 97, 163
Microcoleus lyngbyaceus (filamentous cyanobacterium), 19
Micropogonias undulatus (croaker fish), 121, 124, 125, 126
microscopic organisms, 6, 7, 9, 10–15; archaea, 15–16, 17. *See also* bacteria; eukaryotic microbes

midges (no-see-ums or gnats), 56, 73, 76, 164
mildew, 19–20
mink (*Neovison vison*), 98
minnow, sheepshead (*Cyprinodon variegatus*), 81
Miocene epoch, 34
mixotrophs, 24
Mnemiopsis leidyi (Leidy's comb jelly), 107
molds, 19–20
moles, 8, 97
Molgula (sea grape tunicate), 113–14
mollusks (Mollusca), 7; air-breathing, 56; cephalopods, 117–18, 167; larvae, 108; mud-dwelling, 58–65; nudibranchs, 63, 111; as prey, 141, 152; sand-dwelling, 132–37; shell-less, 58–59, 63, 111. *See also* bivalve mollusks; gastropods
moon snail (*Neverita duplicata*), 133–34, 140
morning glory, beach (*Ipomoea imperati*), 48
Morone saxatilis (striped bass), 126, 175
mosquitoes, 56, 73, 75–77, 164, 175, 176
moss animals, 112. *See also* bryozoans
mouse, oldfield (*Peromyscus polionotus*), 97
mud crabs, 63, 162
mud-dwelling creatures, 49–72; crustaceans, 65–72; macrofauna, 53–57; meiofauna, 51–53, 56, 59, 69–70; mollusks, 58–65
mudflats: at Barn Creek, 50; benthic algae in, 28–29, 170; cordgrass die-off resulting in, 175, 181, 188; plants, 48. *See also* mud-dwelling creatures
mud snail (*Ilyanassa obsoleta*), 29, 59–60, 163, 164–65
Mugil cephalus (mullet), 29, 94–95, 121, 122, 125, 128
Muhlenbergia capillaris (sweetgrass or muhly grass), 44
muhly grass (*Muhlenbergia capillaris*), 44
Mulinia lateralis (surf clam), 135, 136
mullet (*Mugil cephalus*), 29, 94–95, 121, 122, 125, 128
mummichog killifish (*Fundulus heteroclitus*), 79–80, 94. *See also* killifish
mushrooms, 19–20
mussels: as bivalves, 58; fecal material deposited by, 63–64, 162; as filter feeders, 26; food resources in marsh, 169; in fouling communities, 25; as prey, 60, 152, 162; ribbed mussel, 8–9, 63–64, 87, 162, 164. *See also* bivalve

mollusks
Mycteria americana (American wood stork), 93
Myrica cerifera (wax myrtle), 6, 41, 88, 89, 97

Nannygoat Beach, xvi, 30, 47, 122, 132, 146
Navicula (benthic pennate diatom), 29
needlefish, northern (*Strongylura marina*), 129
Negaprion brevirostris (lemon shark), 130
nekton, 115–31; decapods, 118–20; defined, 115; fish, 120–30; marine mammals, 130–31; squid, 115–18
nematocysts, 102, 104, 105, 106, 110
nematodes, 51–52
Nematostella vectensis (starlet sea anemone), 56–57
Nemopilema nomurai (Nomura's jellyfish), 103, 104
Neotoma floridana (eastern wood rat), 97
Neovison vison (American mink), 98
Nephtys (red-lined polychaete worm), 54
Neverita duplicata (moon snail), 133–34, 140
New England salt marshes, xii, 2, 44, 175–76, 181
Nitzschia (benthic pennate diatom), 29
Nitzschia paradoxa (gliding benthic pennate diatom), 29
Noctiluca scintillans (bioluminescent dinoflagellate), 21
nudibranch, 63, 111
Numenius phaeopus (whimbrel), 153
nutrient recycling, 24
Nycticorax nycticorax (black-crowned night heron), 91

octocoral, 110–11
Octopus vulgaris (Atlantic octopus), 58–59, 117
Ocypode quadrata (ghost crab). *See* ghost crab
Odocoileus virginianus (white-tailed deer), 2, 40, 43, 86, 99
Odum, Eugene, ix, 10, 11, 13, 165
oligochaete worms, 53
Oliva sayana (lettered olive snail), 135
Ophiothrix angulata (angular brittle star), 144
Opsanus tau (oyster toadfish), 128
Opuntia humifusa (prickly pear cactus), 43, 45
Opuntia pusilla (creeping cactus), 45–46
orach (*Atriplex cristata*), 46

Orchelimum fidicinium (salt marsh grasshopper), 74, 160
Orchestia grillus (beach flea amphipod), 138
Ordovician period, 34, 117, 143
organic chemical pollutants, 179–80
Ortalis vetula (chachalaca), 95–96
Oryzomys palustris (marsh rice rat), 86, 96–97, 163
osprey (*Pandion haliaetus*), 94–95, 193
otters, 98–99
Ovalipes ocellatus (lady crab), 120
oystercatcher (*Haematopus palliatus*), 152
oyster drill snail (*Urosalpinx cinerea*), 60
oyster reefs, 6; animals living around, 91, 128, 152; *Cliona* demolition of, 110; small animals sheltered by, 61, 63, 66, 67, 71, 81, 144
oysters, 58, 60–63; bryozoans growing on, 112; *Cliona* sponges on, 110; as filter feeders, 26; food in marsh, 169; at high tide, 162; parasitic infection of, 62, 185–86; pea crabs in, 139, 142; as prey, 60, 61, 63, 71, 152. *See also* bivalve mollusks
oyster toadfish (*Opsanus tau*), 128

Pagurus longicarpus (dwarf hermit crab), 140
Pagurus pollicaris (flat-clawed hermit crab), 140
painted bunting (*Passerina ciris*), 89, 193
Paisochelifer (salt marsh pseudoscorpion), 74
Paleomonetes pugio (grass shrimp), 56, 66–67, 80, 118, 124, 162
palmetto, 6, 42, 44
Pandion haliaetus (osprey), 94–95, 193
Panicum amarum (panic grass), 48
Panopeus herbstii (black-fingered mud crab), 71
Pantherophis (rat snakes), 83
Paralichthys dentatus (summer flounder), 127
Paralichthys lethostigma (southern flounder), 127
Parvocalanus crassirostris (coastal copepod), 107
Passerina ciris (painted bunting), 89, 193
Passerina cyanea (indigo bunting), 89
pea crabs, 139, 142
Pelagia noctiluca (mauve stinger jellyfish), 103
Pelecanus occidentalis (brown pelican), 154
Pelegrina tillandsiae (Spanish moss spider), 43
pelicans, 6, 150, 154
penaeid shrimp, 66, 67, 118–19

pennate diatoms, 27, 28–29
pennywort (*Hydrocotyle bonariensis*), 48
perch, silver (*Bairdiella chrysoura*), 122, 124
Perkinsus marinus (protist parasite of oysters), 185–86
Peromyscus polionotus (oldfield mouse), 97
Persephona mediterranea (purse crab), 141
pesticides, 94, 95, 179, 180
Petricolaria pholadiformis (false angel wing clam), 136
Pfiesteria piscicida (fish-killing toxic dinoflagellate), 183
Phalacrocorax auritus (double-crested cormorant), 154
Phoebis sennae (cloudless sulfur butterfly), 78
photosynthesis: by algae, 19, 24–25, 33; C-3 pathway, 34, 160, 168; C-4 pathway, 34, 35, 39, 159–60, 168; carbon dioxide fixation, 18; by dinoflagellates, 21; water loss during, 161
Physalia physalis (Portuguese man-of-war), 106
phytoflagellates, 27–28, 29, 30
phytoplankton, 13, 21, 25–28, 136, 166, 168–70, 178
pines (*Pinus* sp.) 32–33, 42, 44, 91, 93, 177–78
Pinnixa (small commensal crab), 142
pipefish (*Syngnathus*), 129
plankton, xiii; fiddler crab larvae and, 71; filter feeders and, 60, 63–64; menhaden's diet of, 6; role of microbes in, 10, 11–12; shrimp and, 67. *See also* phytoplankton; zooplankton
plankton nets, 101, 107–8, 200–202
plant detritus, 2; eaten by mullet, 128; microbially enriched, 11, 13, 24, 26, 76–77, 137. *See also* cordgrass-detritus food web
plant hoppers, 73, 75, 76, 79–80
Plecia nearctica (lovebug fly), 77–78
Plegadis falcinellus (glossy ibis), 93
Pleistocene epoch, 1–2, 34
Pleurosigma (pennate diatom), 29
plovers, 8, 152
Pluvialis squatarola (black-bellied plover), 152
Podiceps auritus (horned grebe), 155
Poecilia latipinna (sailfin molly), 81
pogies (menhaden, *Brevoortia tyrannus*), 6, 123

Pogonias cromis (black drum), 121, 125
polar ice caps, 171, 187
pollution, coastal, 175, 176–80
polychaetes. *See* bristle worms
Polymesoda caroliniana (marsh clam), 65
Polyodontes lupinus (scale worm), 55
Polysiphonia (filamentous red alga), 31
Pomeroy, Lawrence, xi, xii, xiii, 11–14, 26
Portrait of an Island (J. Teal and M. Teal), xii, 96
Portuguese man-of-war (*Physalia physalis*), 106
Prionotus (sea robin), 128
Procyon lotor (raccoon). *See* raccoon
prokaryotes, 15–19. *See also* bacteria
Prokelisia marginata (salt marsh plant hopper), 73
Proterozoic eon, 19
protists, 10, 13, 14; disease-causing, 62; mixotrophic, 24; photosynthetic, 20, 25, 27–28, 29, 30; predatory, 19, 20–24, 25, 26, 51, 101; role in marine ecosystems, 23–24. *See also specific protist species*
Psathyrella ammophila (mushroom-producing beach fungus), 20
Pseudendoclonium submarinum (filamentous green alga), 31
pseudoscorpion (*Paisochelifer*), 74
pulmonate (air breathing) snails, 56, 59, 162

Quercus virginiana (live oak), ix, 6, 32, 42–44, 93, 96, 99
Quiscalus major (boat-tailed grackle), 88–89

raccoon (*Procyon lotor*), 2, 6, 97–98; as predator, 8, 61, 64, 83, 86, 98, 149, 162
radiolarians, 23
Raeta plicatella (channeled duck clam), 136
Rallus longirostris (clapper rail), 6, 64, 86, 87, 156
rats, 86, 90, 96–97, 163
rat snakes (*Pantherophis*), 83
rattlesnakes, 8, 44, 82, 84–85, 97
rays, 129–30
red tides, 21–22, 164, 183
Renilla reniformis (sea pansy), 111
reptiles, 81–85. *See also Alligator mississippiensis*; sea turtles; snakes; turtles

research: on blue crabs, 120, 184–85; on coastal ecosystems and food webs, xii–xiii, 10–14, 164–70; on coastal fish, 126; on coastal pollution, 177–80; on cordgrass marsh, 10–12, 38, 80, 160, 165–70, 176, 181–82; on dermo disease in oysters, 62, 185–86; on harpacticoid copepods, 53; on marine microbiology, 10–14; on marine protists, 22–23; at University of Georgia Marine Institute, ix, xii–xiii, 10–12, 165–70
respiration in marsh plants, 161
respiration rates in seawater, 11–12, 13
Reynolds, R. J., Jr., ix–xii, 10–11, 92, 95, 198
Rhizoclonium (fuzzy filamentous green alga), 31
rhizomes: of beach pennywort, 48; cordgrass growth and, 3–4, 36, 38, 164; eaten by blue crabs, 120; oxygen funneled to, 36, 160
Rhizosolenia (centric diatom), 27
ribbed mussel (*Geukensia demissa*), 8–9, 63–64, 87, 120, 162, 164
river otter (*Lontra canadensis*), 98–99
rodents, 84, 86, 90, 94, 96–97
ruddy turnstone (*Arenaria interpres*), 152
Russian thistle (*Salsola kali*), 47–48
Rynchops niger (black skimmer), 151, 193

Sabal palmetto (common palmetto or cabbage palm), 42
Sabellaria vulgaris (sand tube worm), 55
Saccoglossus kowalevskii (acorn worm), 49, 51, 55
Sagitta (arrow worm, chaetognath), 108
sailfin molly (*Poecilia latipinna*), 81
Salicornia bigelovii (glasswort), 40
Salicornia virginica (glasswort), 40
salinity: of marsh soils, 4, 35, 65; variation in coastal water, 158–59, 184
Salsola kali (Russian thistle, tumbleweed), 47–48
salt barrens, 39
saltbush (*Atriplex cristata*), 46
salt grass (*Distichlis spicata*), 4, 39
salt marsh, 1; elevation zones, 4; intertidal salt marsh of Georgia coast, 2, 3–6; in New England, 2, 44, 175–76; plants of, 36–41; role in nurturing estuarine fish populations, 126; sea-level rise and, 186–89. *See also* cordgrass marshes; estuaries

Index   229

salt marsh grasshopper *(Orchelimum fidicinium),* 74, 160
salt marsh plant hopper *(Prokelisia marginata),* 73
salt stress, 38, 161, 181–82
salt-tolerant plants, 4, 40–41, 44–48, 159
saltwort *(Batis maritima),* 4, 40, 159, 161
sand-dwelling creatures, 132–45; crustaceans, 137–43; echinoderms, 143–45; mollusks, 132–37
sandflats, 35, 64–65, 68, 133–34, 141, 144
sandpipers, 7, 152–54; as predators, 136, 138; on Tybee Island, 193; in winter, 150, 164
sandspur *(Cenchrus tribuloides),* 48
Sapelo Island: Barn Creek, 50; basket making on, 44–45; birds on, 91–96; Cabretta Beach, 81; Chocolate plantation, 63; Hog Hammock, 194; introduction to, ix, 45; live oaks on, 43; on map, xiv; Marsh Landing Dock, viii, ix, xi, 3, 126; Nannygoat Beach, xvi, 30, 47, 122, 132, 146; Raccoon Bluff, 157–58; shellfish canning operation on, 62; *Spartina* salt marsh on, 5; story tradition on, 157–58; Teal's Boardwalk, xii, 165, University of Georgia Marine Institute on, ix–xiii, 8, 10–11, 165–66
sargassum weed *(Sargassum filipendula),* 31, 148
Scalopus aquaticus (eastern mole), 97
Schizothrix calcicola (filamentous cyanobacterium), 19
Sciaenidae (drum family of fishes), 124–26
Sciaenops ocellatus (red drum), 121, 125
Sclerodactyla briareus (hairy sea cucumber), 145
sea bass, black *(Centropristis striata),* 121, 126
sea cucumbers, 143, 145
sea grasses, 35, 146, 148
seagulls. *See* gulls
seahorse *(Hippocampus erectus),* 129
sea islands of Georgia coast: effect of sea-level rise on, 187–88; formation of, 1–2; map, xiv; salt marshes bordering, 32. *See also specific islands*
sea lavender *(Limonium carolinianum),* 41
sea-level rise, 186–89
sea oats dune grass *(Uniola paniculata),* 7, 8, 47, 48
sea oxeye bush *(Borrichia frutescens),* 4, 40–41, 161
sea pansy *(Renilla reniformis),* 111
sea roaches *(Ligia),* 65, 66

sea robin *(Prionotus),* 128
sea rocket *(Cakile edentula),* 46
seasonal cycles, 163–64
sea squirts (tunicates), 49, 51, 109, 113–14
sea trout, 121, 124–25
sea turtles, 8, 81–82, 113, 146–49, 195, 196
sea urchins, 142, 143, 144, 146
sediments: formation of, 23; green color on, 29; lack of sunlight on, 25; marsh fermenters in, 16; microorganisms in, 24; mussels and, 162; oxygen depletion in, 17, 36; rhizomes extended through, 36; sea grasses in, 35; sulfide smell in, 38; tidal flow and, 2, 4. *See also* mud-dwelling creatures
Selene vomer (lookdown fish), 129
Serenoa repens (saw palmetto), 42
Sesarma reticulatum (marsh crab), 71, 175
sessile plants and animals, 2, 6–7, 106, 109–14, 135
Setophaga coronata (yellow-rumped warbler or myrtle warbler), 89
sewage outflows, 175, 176, 177–78
shad *(Alosa sapidissima),* 121, 123–24
sharks, 121, 129, 130
shark teeth, 130, 193
sheepshead fish *(Archosargus probatocephalus),* 121, 126
shellfish and shellfish industry, 26–27, 58, 61–62, 101, 105
shorebirds, 2, 8, 150–56; active period, 163; on North Beach Birding Trail, 193; as predators, 29, 123; in winter, 164. *See also* birds; *and specific shorebird species*
shrimp, 6, 66–68; as bait, 124, 125, 127; burrowing, 67, 137, 138–39; food in marsh, 170; larvae, 108; living among sea whips, 111; as nekton, 115, 118; in oyster reefs, 63; penaeid shrimp, 66, 67, 118–19; as predators, 24, 29, 56; as prey, 150; seasonal cycle, 163. *See also* grass shrimp
Sicyonia brevirostris (brown rock shrimp), 118–19
Sigmodon hispidus (cotton rat), 8, 97, 163
silica, 23, 27, 109
silversides fish *(Membras martinica* and *Menidia),* 122–23
Sinum perspectivum (white baby ear snail), 134

Skeletonema costatum (planktonic chain-forming diatom), 27
skimmers, 8, 151, 156, 193
smooth cordgrass. *See* cordgrass
snails, 58–60; air-breathing, 56, 59; coffee bean snail, 60, 162; cordgrass as food for, 169; living among sea whips, 111; as predators, 50, 59, 60; as prey, 162; sand-dwelling, 132–35; shells used by hermit crabs, 140. *See also* marsh periwinkle snail; mud snail
snakebird or water turkey (*Anhingha anhinga*), 154
snakes, 8, 82, 83, 84, 90, 163, 195
snapping shrimp (*Alpheus heterochaelis*), 67, 125
snowy egret *(Egretta thula)*, 91–92
soft corals, 110–11
soil salinity, 4–6, 35, 39, 161
Solidago sempervirens (seashore goldenrod), 46
southern yellow pines, 42, 44
spadefish, Atlantic (*Chaetodipterus faber*), 129
Spalding, Thomas, 45
Spanish bayonet lily (*Yucca aloifolia*), 45
Spanish moss (*Tillandsia usneoides*), 43
sparrows, 76, 88
Spartina alterniflora (smooth cordgrass). *See* cordgrass
Spartina cynosuroides (giant cordgrass), 38
Spartina patens (salt marsh hay cordgrass), 44, 97
Sphyrna (hammerhead shark), 130
spider crabs (*Libinia*), 105, 141–42
spiders, 43, 52, 73, 74–76; in beach dunes, 77; as prey, 79–80, 87; spotting at night, 163
sponges, 6, 109–10; evolution of, 20; in oyster reefs, 63; as prey, 63, 146; spider crabs and, 142
Sporobolus virginicus (dropseed grass), 39
spurges, 46
squareback crabs, 71, 87, 96, 160, 175–76
squid, 58–59, 115–18, 146
Squilla empusa (mantis shrimp), 68
stable isotope analysis, xiii, 160, 166–69
starfish (sea stars), 7, 25, 143–44
starlet sea anemone (*Nematostella vectensis*), 56–57
Stellifer lanceolatus (star drum), 125, 126

Sterna hirundo (common tern), 151
Sternula albifrons (least tern), 151
stingray (*Dasyatidae*), 129–30
Stomolophus meleagris (cannonball jellyfish), 105
striped bass (*Morone saxatilis*), 126, 175
Strongylura marina (northern needlefish), 129
subtidal estuary, 2, 6–7, 101. *See also* estuaries
succulent plants, 35, 39, 40, 45, 46, 97
sulfate reducing bacteria, 16–18, 36
sulfide: inhibition of plant growth by, 36, 38, 159; mussels smelling of, 64; oxidation of, 17–18, 24; production of, 16–17; used by cordgrass plants, 160
sulfide oxidizing bacteria, 17–18, 24
Superfund sites, 179
swallow, tree (*Tachycineta bicolor*), 88
sweetgrass (*Muhlenbergia capillaris*), 44
Sylvilagus aquaticus (freshwater marsh rabbit), 97
Sylvilagus palustris (tidal marsh rabbit), 97
Symphurus plagiusa (blackcheek tonguefish), 127
Synedra (fouling community diatom), 29
Syngnathus (pipefish), 129
systems ecology, 10–11

Tabanus nigrovittatus (greenhead fly), 56, 76
tabby stone, 62–63, 195
Tachycineta bicolor (tree swallow), 88
Tagelus plebeius (stout razor clam), 137
tanaid (*Hargeria rapax*), 55–56
tarpon (*Megalops atlanticus*), 115, 121, 124
Teal, John: cattle egrets spotting, 92–93; chachalacas noted by, 96; deer noted by, 99; food web concept, xii–xiii, 11, 13, 165; *Life and Death of the Salt Marsh*, xii, 175; *Portrait of an Island* (with M. Teal), xii, 96; on raccoon survival, 98; white ibis noted by, 93
Teal, Mildred, xiii, 98, 99, 175; *Portrait of an Island* (with J. Teal), xii, 96
Tellina (marine clam), 136–37
Tellina alternata (alternate tellin clam), 137
ten-pounder (*Elops saurus*), 124
Terebra dislocata (eastern auger snail), 135
terns, 6, 8, 150, 151, 156, 193
terrapin (marsh turtle), 81–82, 83, 98, 163

Tertiary period, 34, 87
Thalasseus maximus (royal tern), 151
Thalasseus sandvicensis (Sandwich tern), 151
tidal creeks, 2, 4, 5, 6; cordgrass growing along, 32, 36, 38, 175, 176; creekside marsh, 4, 5, 36, 37; dinoflagellate blooms, 21–22; human-made, 175, 176; muds of creek bottoms and banks, 51, 54; tidal flooding of, 11, 38
tidal cycle, animal activity during, 162–63
tidal flooding, 1, 2, 4, 11, 159
Tillandsia usneoides (Spanish moss), 43
toxic chemical pollution, 178–80
Trachinotus carolinus (Florida pompano), 126
Trinectes maculatus (hogchoker flatfish), 127
Tringa flavipes (lesser yellowlegs sandpiper), 153
Tringa melanoleuca (greater yellowlegs sandpiper), 153
Tringa semipalmata (willet), 153–54
trout, sea, 121, 124–25
tumbleweed (*Salsola kali*), 47–48
tunicates, 49, 51, 109, 113–14
Tursiops truncatus (bottlenose dolphins), 6, 115, 123, 130–31, 179–80, 193
turtles, 8, 81–82, 113, 146–49, 195, 196
Tybee Island, xiv, 87, 193

Uca minax (red-jointed fiddler crab), 69
Uca pugilator (sand fiddler crab), 39, 69, 126, 163
Uca pugnax (marsh fiddler crab), 29, 68–69, 164
Uca thayeri (mangrove fiddler crab), 70
Ulva (sea lettuce), 31
Uniola paniculata (sea oats dune grass), 7, 8, 47, 48
University of Georgia Marine Institute, 8, 10, 165; history, ix–xii
Upogebia affinis (coastal mud shrimp), 67
Urosalpinx cinerea (oyster drill snail), 60

vascular plants, 32, 33
vertebrates, 49, 79–100; reptiles, 81–85. *See also* birds; fish; mammals; *and specific species*

viruses, 10, 14–15
vultures, 95

warblers, 43, 89, 194
Wassaw Island, xiv, 146
wax myrtle (*Myrica cerifera*), 6, 41, 88, 89, 97
weather, extreme, 174, 181–82; hurricanes, 157–58, 187
whales, 115, 130–31
wharf crab (*Armases cinereum*), 71, 162
whelks, 58, 132–33, 140
whimbrel (*Numenius phaeopus*), 153
whiting (kingfish, *Menticirrhus*), 121, 124
willet (*Tringa semipalmata*), 153–54
wood stork (*Mycteria americana*), 93, 194
worms: acorn worms, 49, 51, 55; annelids, 53; arrow worms, 108; earthworms, 97; food sources for, 29; fouling communities and, 109; macrofaunal, 53–55; meiofaunal, 51–52; oligochaete worms, 53; as prey, 29, 97, 146, 152. *See also* bristle worms
wrack, beach, 7–8, 10, 32, 133, 137–38, 163
wrens. *See* marsh wrens

yaupon (*Ilex vomitoria*), 42
yellow-rumped warbler (*Setophaga coronata*), 89
Yucca aloifolia (Spanish bayonet lily), 45

Zanthoxylum clava-herculis (Hercules-club), 42
Zaops ostreus (oyster pea crab), 142
zoea (crab larva), 68
zooplankton, 101–8; collection methods, 199–201; crustaceans, 107–8; diversity of, 51, 53; gelatinous, 101–7; invertebrate larvae, 108; metabolic activity, 11–12; as prey, 101, 104, 106, 110, 129, 148; role in marine ecosystems, 6, 26, 115